Applied Mathematical Sciences | Volume 6

W. Freiberger
U. Grenander

A Course in Computational Probability and Statistics

With 35 Illustrations

Springer-Verlag New York · Heidelberg · Berlin
1971

Walter Freiberger
Ulf Grenander
Division of Applied Mathematics
Brown University
Providence, Rhode Island

Library of Congress Catalog Card Number 76-176272.
Printed in the United States of America.

ISBN 0-387-90029-2 Springer-Verlag New York • Heidelberg • Berlin
ISBN 3-540-90029-2 Springer-Verlag Berlin • Heidelberg • New York

PREFACE

This book arose out of a number of different contexts, and numerous persons have contributed to its conception and development.

It had its origin in a project initiated jointly with the IBM Cambridge Scientific Center, particularly with Dr. Rhett Tsao, then of that Center. We are grateful to Mr. Norman Rasmussen, Manager of the IBM Scientific Center Complex, for his initial support.

The work is being carried on at Brown University with generous support from the Office of Computing Activities of the National Science Foundation (grants GJ-174 and GJ-710); we are grateful to Dr. John Lehmann of this Office for his interest and encouragement. Professors Donald McClure and Richard Vitale of the Division of Applied Mathematics at Brown University contributed greatly to the project and taught courses in its spirit. We are indebted to them and to Dr. Tore Dalenius of the University of Stockholm for helpful criticisms of the manuscript.

The final stimulus to the book's completion came from an invitation to teach a course at the IBM European Systems Research Institute at Geneva. We are grateful to Dr. J.F. Blackburn, Director of the Institute, for his invitation, and to him and his wife Beverley for their hospitality.

We are greatly indebted to Mrs. Katrina Avery for her splendid secretarial and editorial work on the manuscript.

TABLE OF CONTENTS

INTRODUCTION

The purpose of this book is to present an attitude. It has been designed with the aim of making students and perhaps also faculty aware of some of the consequences of modern computing technology for probability theory and mathematical statistics. Not only the increased speed and memory of modern computers are relevant here; of at least equal importance to our subject are the versatile input-output devices and the existence of interactive time-sharing systems and of powerful programming languages. Of the last-mentioned, we have found APL most useful for our purposes.

The work described in these notes was initiated because we felt the time was ripe for a systematic exploitation of modern computing techniques in mathematical statistics and applied probability. Model building, for instance in applied probability, is very different today from what it was in pre-computer days, although this change has not yet fully penetrated to the textbook level. This course is being presented to remedy this situation to some degree; through it, we hope, students will become aware of how computers have increased their freedom of choice of mathematical models and liberated them from the restraints imposed by traditional mathematical techniques.

The project which led to this set of lecture notes is a continuation, although in different form, of an activity organized several years ago at the University of Stockholm. The activity was intended as a complement to the already existing program for training graduate students in mathematical statistics and operations research. It was felt that the students received a well-balanced education in mathematical theory but that something else was lacking: they were not trained for solving real-life problems in raw and unpolished form (in which they usually appear), far removed from that of pure and idealized textbook problems. In particular, little attention was given to the first, difficult stages of problem solving, namely the building of models, the collection of data, and the crucial relation between the mathematical, formal apparatus and the real-life phenomena under study.

To give students an opportunity for acquiring a more realistic orientation and critical attitude towards data, they were exposed to real problems chosen from many fields in industry, government and research. With the help of advice from a teacher or from older and more experienced students, they were asked to study a problem, formulate their own model, scrutinize data and present an analysis to the person or agency from whom the problem had originated. The results were later discussed in laboratory sessions, often before the analysis was completed, in order to get suggestions and ideas.

It became obvious in the course of this experiment that a major defect in the conventional training of graduate students was the lack of attention paid to the role of computers in the applications of mathematical statistics. Students have often to pass through a more or less painful stage in which they reorient themselves in order to learn what is computationally feasible as distinct from analytically possible. It is desirable that this transition be made easier, quicker and more complete. Although most students will have had some exposure to the computer, they may be inexperienced in its use in, say, applied probability. This will affect their ability to formulate realistic models, forcing them to choose analytically tractable models rather than those which best fit the problem.

The purpose of the present project is to equip students in probability and statistics better for work on real problems. The emphasis in the latter effort is on model building as influenced by computer science.

A growing number of practicing statisticians are today aware of the need to exploit more fully the capability of computers in statistics. This is particularly true in applied statistics, for instance in data analysis, where some research workers have emphasized this for several years. Less attention has been paid to the computational side of probability theory, although the need for a computational reorientation also exists in this subject. We therefore chose to concentrate our efforts on <u>probability theory and applied probability as well as on statistics</u>.

We divided our work into several chapters. Each chapter represents some concept or technique or relation for which a sufficiently rich mathematical structure has been developed and in which, at the same time, the impact of computer science can be expected to be substantial. The chapters together will cover only a part of mathematical statistics, although, we hope, an important one. We are particularly interested in areas in which <u>the interaction between the analytical and computational approach</u>

is strong. This will usually only be the case where the analytical technique has been extended so far that further extension seems possible or worthwhile only through computer use, and makes it necessary that students possess a certain degree of mathematical sophistication. A course designed in such a way should be distinguished from one aiming only at teaching the computer implementation of standard statistical techniques and the writing of statistical programs. A course of the latter type is certainly useful and necessary, but the present project is more ambitious in scope and perhaps also more difficult to accomplish in the classroom. Little, in fact, seems to have been done in the direction described here. We had originally hoped to be able to use some already developed material, but there is disappointingly little available.

The prerequisites for the course are familiarity with the elements of probability and statistics, calculus and linear algebra. It will be assumed that students have some programming experience; most computing in the course will be based on APL which is, from the point of view of our computing tasks, by far the most suitable currently available interactive programming language.

The APL programs in the book should not be interpreted as forming a comprehensive, debugged program library (see section 7.1 in this context). They are presented only to illustrate our approach to computational probability and statistics and undoubtedly contain several mistakes.

Since the degree of mathematical sophistication is expected to vary a good deal among students, more advanced sections are starred, and it is suggested that they be read only by those who feel they have sufficient mathematical background. These sections should be discussed in class, without detailed proofs being given; instead, their interpretation and practical consequences should be discussed by the lecturer.

We strongly recommend that students be encouraged to complete the assignments to help them in the development of a real understanding of the material. The extent and size of the assignments will depend in part upon the computational facilities available during the course.

For further reading, and to find more advanced developments of some of the subjects covered here, we recommend the series of reports published under the NSF-sponsored "Computational Probability and Statistics" project at Brown University, the titles of which are listed in the References.

The curves in the book were produced by a TSP plotting system on-line with a DATEL terminal, operating under APL/360.

CHAPTER 1:

RANDOMNESS

1.1 Fundamentals.

The concept of randomness is fundamental in probability theory and statistics, but also most controversial. Among the many interpretations of terms like probability, likelihood, etc., we shall consider two in this course: the usual frequency approach (in this chapter) and the Bayesian one (in chapter 6: "Decision problems").

One should actually not speak of a single frequency approach, since there are several variations of it. That most commonly adopted in the textbook literature is to start from the idea of a random experiment and carry out the mathematical formalization as follows.

Starting from a sample space X that may be completely unstructured, one views the outcome of the random experiment E as a realization of a stochastic variable x described by a probability measure P given on X. The pure mathematician is wont to phrase this as follows: the value of the probability P(S) should be defined for any subset $S \subset X$ belonging to a well-defined σ-algebra of subsets of the sample space. (We shall not, however, go into the measure-theoretical aspects in this course.) This is the mathematical model: the transition to phenomena of the real world is effected through a heuristic principle: the frequency interpretation of probabilities. If the experiment E is repeated n times independently and under equal conditions, then the relative frequency f/n should be close to P(S), where f is the absolute frequency (number of times we get a value $x \in S$), if the sample size, n, is large enough.

While the idea behind this ancient principle is quite appealing, the above formulation is not quite clear on three points:

a) what is meant by "independently"?

b) how should one interpret the phrase "under equal conditions"?

c) how large is "large enough"?

It has often been argued that this sort of vagueness must always be expected when any mathematical model is interpreted in terms of the physical world. For example, when we use Euclidean plane geometry to describe and analyze measurements of length, angles and areas, we meet the same sort of difficulty when trying to relate notions like points, lines and areas to physical data. We continue to use the model only as long as no logical inconsistency is found or no serious discrepancy between model and data has been established empirically. This pragmatic attitude has been widely accepted, but doubts have been voiced by critics who claim that a more profound analysis is possible. To understand better how this can be done we shall take a look at the manner in which we actually use the model.

Simplifying drastically, we could say that from the above point of view probability and mathematical statistics are the study of bounded measures. While such a statement would undoubtedly describe much research activity in these fields quite accurately, it is a superficial point of view and of little help when we want to discuss the relation between theory and its application.

Randomness enters on three different levels. The _first_ can be exemplified by a set of measurements of some physical constant like the speed of light. Here we would think of the variation encountered among the data as caused by imperfections in the experimental arrangement, imperfections which could, at least in principle, be eliminated or reduced by building better equipment or using a more efficient design for the experiment. We describe this variation in probabilistic terms, but probability plays here only a marginal role. On the _second_ level randomness plays a more fundamental role. Assume that we measure the toxicity of a drug and use guinea pigs in the experiment. We would have to consider the apparent randomness caused by the biological variation in the population of animals used. We would always expect such variation, although its form and extent might vary between different populations. In a well-designed experiment we would like to be able to make precise statements about this variation; to eliminate it entirely does not seem possible. Most of experimental statistics falls into this category. To illustrate the _third_ level, let us think of a Monte Carlo experiment in which we try to find the properties of a statistical technique by applying it to artificial data and studying the result, or in which a probabilistic limit theorem is examined by simulating it for large samples. Here we _need_ randomness, and try to generate data in some way so that they have properties we would expect from random sequences. Another example bringing out this feature perhaps even more clearly occurs in the design of experiments when we inject randomness intentionally into the experimental arrangement. Here randomness is desirable and necessary.

In recent years simulation of probabilistic phenomena has become a useful and often-applied tool, especially in the area of operations research. Almost all of this work is being done on computers and it is obvious that this fact has greatly influenced our interest in computer generation of random sequences. There is a close relation between certain results in analytical probability theory, some of which have been known for many years, and the more algorithmically-oriented computer studies carried out recently. This relation will be examined in the following sections.

1.2 Random Number Generation.

Let us now turn to the question of how one could generate randomness on a computer.

We shall try to construct an algorithm that generates numbers $x_1, x_2, x_3, \ldots$ in the interval (0,1) such that the sequence $\{x_i\}$ behaves as one would expect from a sample from the rectangular (uniform) distribution over the same interval (0,1).

What properties should we ask for? Let us mention a few. We know from the law of large numbers that the average

$$\bar{x} = \frac{1}{n}\sum_{1}^{n} x_i \tag{1.2.1}$$

should tend in probability to 1/2 and we would require $\bar{x}$ to be close to 1/2 in some sense that has to be left somewhat vague at present. Similarly, the empirical variance

$$s_n^2 = \frac{1}{n}\sum_{1}^{n} (x_i - \bar{x})^2 \tag{1.2.2}$$

should tend to 1/12 in probability. We can think of other similar properties; let us mention some others. We certainly want the sample $(x_1, x_2, \ldots, x_n)$ to have an approximately uniform distribution over (0,1), e.g. in the following sense. We divide the interval into the m intervals $I_\nu = (\frac{\nu-1}{m}, \frac{\nu}{m})$, $\nu = 1,2,\ldots,m$, where $m \ll n$, and call N_ν the number of x_i that fall in I_ν; then we would expect N_ν/n to be close to $1/m$ for $\nu = 1,2,\ldots,m$.

If we want the algorithm to have a simple form and still generate long (or even infinite) sequences, it is almost necessary that it be recursive. The simplest would be to set

(1.2.3) $$x_{i+1} = f(x_i) , \quad i = 1,2,...$$

where f is a predetermined function taking values in (0,1) (one could also let f depend on more than one of the x's preceding x_{i+1}). It must also be simple to compute, since (1.2.3) will be repeated many times. Starting from a value $x_i \varepsilon (0,1)$ and applying (1.2.3) recursively, we get a sequence of unlimited length.

To start with, let us choose f as

(1.2.4) $$f(x) = \frac{1}{2} + 4(x - \frac{1}{2})^3 ;$$

it will soon become obvious why we have chosen a function of this form (which is graphed in figure 1).

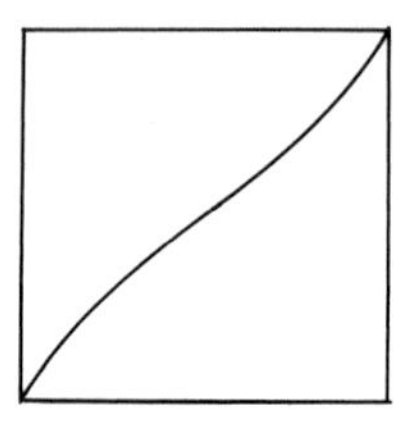

Figure 1.

Assignment: Write a program in APL to generate a series of length n. Calculate $\bar{x}$, s_n^2 and the values of N_ν/n for some suitable value of n. Work with small values of n, say 20.

Let us run the program and note what happens.

To avoid the clearly non-random clustering observed in the above example, let us use

(1.2.5) $$f(x) = \{x + a\} ,$$

starting with $x_1 = 0$, where a is some fixed number in (0,1). Here {x} stands for the fractional part of x or, in APL, x-⌊x.

Let us now write the same program as before and execute it at the terminal. What behavior do we observe? What about special values of a?

Let us now do the same with

$$f(x) = \{2x\} \tag{1.2.6}$$

Let us also try larger values of n, say 100, and see what happens then. We may also be interested in replacing (1.2.6) by $f(x) = \{Mx\}$, M = any natural number, or by more "chaotic" functions, such as

$$f(x) = \begin{cases} 2x & 0 \le x \le \frac{1}{3} \\ 2 - 3x & \frac{1}{3} \le x \le \frac{2}{3} \\ 3x - 2 & \frac{2}{3} \le x \le 1 \end{cases} \tag{1.2.7}$$

There may be other suitable choices of f with which to run your program. What can we say about the behavior of the sequence generated?

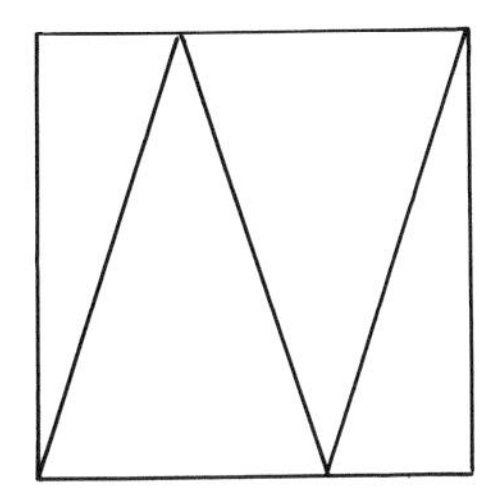

Figure 2

The reason why (1.2.4) failed is obvious. Indeed, by looking at the graph of the function plotted in figure 2, it is clear that if $1/2 < x_i < 1$ then $1/2 < x_{i+1} < x_i$, and that the sequence converges to 1/2; the same is true if $0 < x_i < 1/2$. The three values x = 0, 1/2, 1 play the role of fixed points: the middle one is stable and the other two are unstable. We would, of course, like to avoid fixed points. The function is not sufficiently irregular to generate the chaotic behavior we want. Actually, for any f that is continuous and maps [0,1] onto itself there is at least one fixed point. This leads us to introduce discontinuous functions.

Consider instead the function (1.2.5); it has a discontinuity at x = 1-a (plot the function!). The following beautiful result due to Herman Weyl illustrates the situation well.

Theorem. For any irrational $a \in (0,1)$ the sequence $x_1, x_2, x_3, \ldots$ with $x_i = \{ia\}$ is equidistributed over $(0,1)$ (this relates to additive congruence generators; see below).

Note: by equidistribution in this context - for sequences of numbers and not for stochastic variables - is meant the following. For any a,b with $0 \leq a < b \leq 1$ we should have

(1.2.8) $$\lim_{n\to\infty} \left[\frac{1}{n} \times \text{number } \{x_i: a \leq x_i \leq b;\ i=1,2,\ldots,n\}\right] = b - a$$

*Proof of theorem (asterisks denote the beginning and end of an advanced section). Consider the trigonometric sums

(1.2.9) $$T_k = \frac{1}{n} \sum_{j=1}^{n} e^{2\pi i k\{ja\}} = \frac{1}{n} \sum_{j=1}^{n} e^{2\pi ikja}$$

We have $T_0 = 1$ and for $k \neq 0$ by summing the geometric series

(1.2.10) $$|T_k| \leq \frac{1}{n} \frac{1}{\sin \pi ka}$$

Since a is irrational, the value ka is not an integer, so that $\sin \pi ka \neq 0$. We get from (1.2.10)

(1.2.11) $$\lim_{n\to\infty} T_k = 0$$

This implies that for any trigonometric polynomial $P(x) = \sum_k a_k e^{2\pi ikx}$ we have

(1.2.12) $$\lim_{n\to\infty} \frac{1}{n} \sum_{j=1}^{n} P(x_j) = \int_0^1 P(x)\, dx$$

Since any continuous function on (0,1) can be uniformly approximated by trigonometric polynomials, (1.2.12) holds when P is an arbitrary continuous function. Introduce, for $\varepsilon > 0$, the two stepwise linear functions f^+ and f^- (see figure 3). They are con-

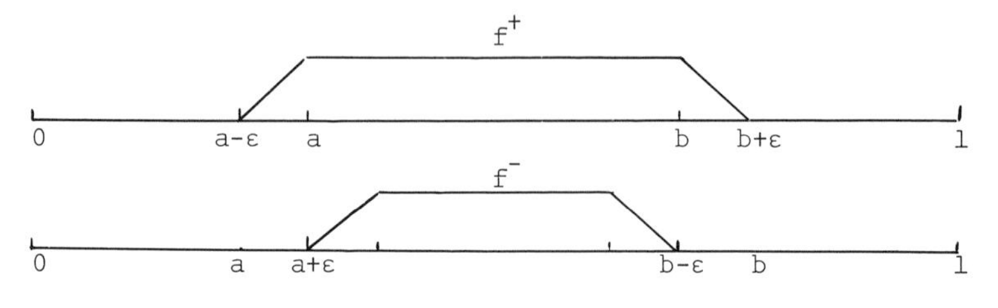

Figure 3

tinuous so that (1.2.12) holds for them. But the quantity in (1.2.8) is included between the two limits we just got; this proves the theorem.*

Note: this theorem does not tell us for which values of x_1 we get equidistribution, only that equidistribution is the typical case.

There is also the following result:

Theorem. For almost all initial values $x_1 \in (0,1)$, the sequence generated by (1.2.6) is equidistributed (this relates to multiplicative congruence generators; see below).

*Proof. The mapping Tx, where

$$T: x \to \{2x\} \tag{1.2.13}$$

of (0,1) onto itself preserves Lebesgue measure. Indeed, let f(x) be an arbitrary periodic (with period 1) bounded and measurable function. Then

$$\int_0^1 f(Tx)dx = \int_0^1 f(2x)dx = \frac{1}{2}\int_0^1 f(y)dy = \int_0^1 f(y)dy \tag{1.2.14}$$

Taking f as the indicator function of an arbitrary subinterval of (0,1) proves that T is measure-preserving in the sense that $m(T^{-1}S) = m(S)$ for any Borel set $S \in [0,1]$. This implies that if x_1 is given a uniform distribution over (0,1) the sequence x_i (now consisting of stochastic variables) forms a stationary stochastic process and we can apply Birkhoff's individual ergodic theorem (see ref. 13, p. 105) and find that for any $g \in L_1(0,1)$

$$\hat{g} = \lim_{n\to\infty} \frac{1}{n} \sum_{j=1}^{n} g(x_n) \tag{1.2.15}$$

exists for almost all choices of x_1. It remains only to prove that the limit $\hat{g}$ is essentially constant; it must then be equal to

$$\bar{g} = \int_0^1 g(x)dx \qquad \text{(almost certainly)} \tag{1.2.16}$$

which would prove the equidistribution. To prove that g is constant it is enough to prove metric transitivity (ergodicity) of T (see ref. 13, p. 105). Let I be an invariant set with the indicator function I(x):

$$I(Tx) = I(\{2x\}) = I(x), \quad \text{all } x \in (0,1) \tag{1.2.17}$$

Form the kth Fourier coefficient a_k of $I(x)$:

$$a_k = \int_0^1 e^{2\pi ikx} I(x)\,dx \tag{1.2.18}$$

Then we get from (1.2.17) and the periodicity of the trigonometric function

$$a_{2k} = \int_0^1 e^{4\pi ikx} I(x)dx = \int_0^1 e^{2\pi ikx} I(\{2x\})dx = \frac{1}{2}\int_0^1 e^{2\pi ikx} I(\{x\})dx \tag{1.2.19}$$

$$= \int_0^1 e^{2\pi ikx} I(x)dx = a_k$$

Therefore $a_k = a_{2k} = a_{4k}$, etc. According to the **Riemann-Lebesgue lemma (see ref. 14,** p. 469), the Fourier coefficients $a_k \to 0$ as $h \to \infty$ so that $a_{\pm 1} = a_{\pm 2} = \ldots = 0$ and I is identically constant almost certainly, which completes the proof.*

While these considerations help to understand what happened computationally, it is clear that they leave out one aspect: we work, on a digital machine, not with all real numbers but with only a finite subset of them. Let us look into some consequences of this consideration.

We noticed that there were certain initial values x_1 that led to degeneracy when we generated the sequence. If f stands for the mapping used, it is clear that solutions of the equation

$$f(x) = x, \quad 0 \leq x \leq 1 , \tag{1.2.20}$$

the fixed points of f, used as initial values, will give $x_1 = x_2 = x_3 = \ldots$. Similarly, if f^p stands for the pth iterate of f, then the fixed points of f^p will give a periodic sequence with period at most p.

On the other hand, the values of x at our disposal in the machine are only finite in number and our results depend upon the representation they are given in the computer. Assume they are represented as binary fractions with d digits. If $f(x) = \{2x\}$, for example, this means that when $x > 1/2$, we will lose one digit when computing $f(x)$ in floating-point arithmetic. This will, when repeated several times, lead to degeneracy of the sequence.

Note that, if we have access to integer arithmetic, we can always work with x as the natural numbers (in the integer mode). When performing a multiplication $M \cdot x$ we will sometimes encounter overflow, but this will lead to a loss of the first few digits only; we actually do the reduction modulo K where K-1 is the largest integer the machine accepts. This saves us from some of the troubles observed above. The numbers generated would then be of the form X/K, i.e. multiples of 1/K.

Returning to the fixed points of T^p, we note that they will be critical computationally; or, more precisely, those numbers will be so which the machine can handle and to which the fixed points are rounded off (with a wide interpretation of this term). This explains some peculiarities we may have encountered computing at the terminal.

Obviously all methods of the type described above will be periodic because of the digital nature of the computations and the finite memory size. There are ways of ensuring that this period is very large; more about this later.

At this point it is natural to ask the following question. We have studied some deterministic sequences and shown that some do exhibit some of the properties one would expect of a truly random sequence. The mathematical explanation of this behavior is not very hard to follow. On the other hand we know, or believe, that some sequences found in nature really should be described as random. This fact has often been used to get random sequences, for instance by using radioactive decay or cosmic radiation, or by mechanical devices such as dice or roulette wheels. Is it possible to give a mathematical "explanation" of this often-observed randomness?

This is one of the fundamental problems in statistical mechanics. We shall take a look at it for the case of a conservative mechanical system.

To a reader familiar with rational mechanics it is well known that such a system can be described in canonical coordinates $q_1, q_2, \ldots, q_f$ and the corresponding canonical moments $p_1, p_2, \ldots, p_f$ by using a function, the Hamiltonian $H = H(q_1, \ldots, q_f, p_1, \ldots, p_f)$. Here f is the number of degrees of freedom. The equations of motion are

$$
\begin{aligned}
dq_\nu/dt &= \partial H/\partial p_\nu \\
dp_\nu/dt &= -\,\partial H/\partial q_\nu
\end{aligned}
\qquad \nu = 1,2,\ldots,f \tag{1.2.21}
$$

Denote by T_t the transformation in the phase space R^{2f} corresponding to a motion of t units of time. The group $\{T_t\}$ of such transformations has the very important property that <u>it leaves Lebesgue volume invariant</u> (Liouville's theorem), which has direct consequences for the apparent randomness.

To prove this, it is enough to show that the Jacobian J is constant and equal to 1. Differentiate J with respect to time:

$$(1.2.22)\qquad \frac{dJ}{dt} = \frac{d}{dt}\frac{\partial(x_1\ldots x_{2f})}{\partial(y_1\ldots y_{2f})} = \sum_{k=1}^{2f}\frac{\partial(x_1,\ldots\dot{x}_k,\ldots x_{2f})}{\partial(y_1,\ldots\ldots\ldots y_{2f})} = \sum_{k=1}^{2f} J_k$$

say, where we have used the notation $x_1 = q_1$, $x_2 = q_2,\ldots,x_{f+1} = p_1$, $x_{f+2} = p_2$, etc., for the final position and y for the initial position in phase space. Then, using (1.2.21) and simple properties of determinants, we get

$$(1.2.23)\qquad J_k = \sum_{\nu=1}^{2f}\frac{\partial(x_1,\ldots x_{k-1},x_\nu,x_{k+1}\ldots x_{2f})}{\partial(y_1,\ldots\ldots\ldots\ldots\ldots\ldots y_{2f})}\frac{\partial\dot{x}_k}{\partial x_\nu} = J\frac{\partial\dot{x}_k}{\partial x_k}$$

and

$$(1.2.24)\qquad \frac{dJ}{dt} = J\sum_{k=1}^{2f}\frac{\partial\dot{x}_k}{\partial x_k} = J\Big[\sum_{k=1}^{f} + \sum_{f+1}^{2f}\Big]\frac{\partial\dot{x}_k}{\partial x_k} = J\Big[\sum_{k=1}^{f}\frac{\partial\dot{q}_k}{\partial q_k} + \sum_{k=1}^{f}\frac{\partial\dot{p}_k}{\partial p_k}\Big]$$

$$= J\Big[\sum_{k=1}^{f}\Big(\frac{\partial^2 H}{\partial p_k\partial q_k} - \frac{\partial^2 H}{\partial q_k\partial p_k}\Big)\Big] = 0$$

This proves that Lebesgue volume is invariant. We can now apply ergodic theory to any integrable function defined in phase space and can make statements like those made about some pseudo-random sequences earlier in this section. We can certainly do this if we have ergodicity, and this condition may be difficult to guarantee. At least, however, this approach helps us to understand why the apparent randomness is found so often in mechanical and other physical systems.*

We should note that this is not a general proof of randomness for mechanical systems; we would first have to look for invariants like energy, total moments, etc., and then establish ergodicity for the motion restricted to a manifold characterized by these invariants. It would be dangerous to treat series obtained in this way as random without a detailed analysis, and the same holds <u>a priori</u> for other physically generated "random" sequences.

To look into this question computationally, let us consider a mass point moving without friction inside a square with elastic collisions with the walls. Denoting the coordinates at time t by x(t), y(t), we can consider the motion as a modification of rectilinear motion (with no walls).

Let

$$\hat{x}(t) = x(0) + t\cdot v_x$$
(1.2.25)
$$\hat{y}(t) = y(0) + t\cdot v_y$$

and describe the motion by

$$x(t) = \text{per}(\hat{x}(t))$$
(1.2.26)
$$y(t) = \text{per}(\hat{y}(t))$$

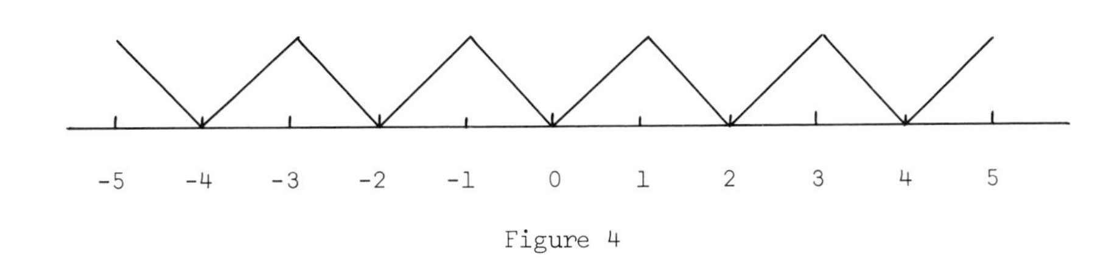

Figure 4

where the function per is periodic and looks as is shown in figure 4. Let us count how many times the point passes through a little square

(1.2.27) $S_{ij} = \{ih \leq x < (i+1)h;\ jh \leq y < (j+1)h\}$, $i,j=0,1,2,\ldots N-1$, $h = 1/n$

during a given duration. Call this number N_{ij}. An APL program called MOTION is given in Appendix 1 (figure 1.1), together with results displaying the numbers N_{ij} in a table with the convention that if $N_{ij} = 0$ this is represented by a blank and if $N_{ij} > 9$ an asterisk is printed.

Using 300 time steps we get figure 1.2, with velocity components XVEL = .05 and YVEL = .1; this shows a very uneven distribution over the unit square. Changing XVEL to 1/30, we get the trajectory in figure 1.3, rather more spread out. Finally, XVEL = .11 results in figure 1.4, where the density displays a coarse-grained uniformity. Note that the ratios of the velocity components were 2, 3 and 10/11 respectively. It is possible to explain the behavior observed above by the following theorem.

*Theorem. Call the ratio YVEL/XVEL = ρ. Then a sum of the form

$$\frac{1}{A}\sum_{t=1}^{A} I(x(t),y(t))$$

where I is the indicator function of some rectangle $R = \{x,y \mid a_1 \leq x \leq a_2,\ b_1 \leq y \leq b_2\}$, tends to the area of R if ρ is irrational.

Proof. Let $f(x,y)$ be an arbitrary continuous function in the unit square and consider the sums

$$S_A(f) = \frac{1}{A} \sum_{t=1}^{A} f(x(t),y(t))$$

We shall show that

$$\lim_{A\to\infty} S_A(f) = \int_0^1 \int_0^1 f(x,y)\, dx\, dy$$

Note that, using (1.2.26), we can write $f(x,y) = g(\hat{x},\hat{y})$ where g is periodic in both variables with period 2. Let us look at the special case

$$g(\hat{x},\hat{y}) = e^{\pi i(k\hat{x} + \ell\hat{y})}$$

where k and ℓ are integers. Then the sums can be written as

$$\frac{1}{A} \sum_{t=1}^{A} \exp[\pi i(kx(0)+ktv_x + \ell y(0)+\ell tv_y)] =$$

$$= \frac{1}{A} \exp[\pi i(kx(0)+kv_x + \ell y(0)+\ell v_y)] \frac{1 - \exp[iA\pi(kv_x+\ell v_y)}{1 - \exp[i(kv_x+\ell v_y)]} \to 0 \text{ as } A \to \infty$$

if $kv_x + \ell v_y \not\equiv 0 \bmod 2$ and with the limit

$$\exp[\pi i(kx(0)+\ell y(0))]$$

if $kv_x + \ell v_y \equiv 0 \bmod 2$. But if ρ is irrational the congruence is true only if $k = \ell = 0$ so that the limit is then equal to 1. In both cases the limit can be written as

$$\frac{1}{4}\frac{1}{4}\int_0^2 \int_0^2 g(\hat{x},\hat{y})\, d\hat{x}\, d\hat{y} = \int_0^1 \int_0^1 f(x,y)\, dx\, dy$$

Hence the theorem is true in this particular case as well as when g is any trigonometric polynomial with periods 2. The result now follows by the same approximation as was used in the proof of Weyl's theorem above.*

It is clear that the measure-preserving nature of our chaos-increasing transformations is fundamental. When we operate digitally, as in the computer, measure = counting measure, in that measure-preserving is equivalent to the transformation being one-one. Consider a mixed congruence generator with the mapping $T: x \to y$

(1.2.28) $$y \equiv a\cdot x+b \pmod N$$

In order that this transformation be one-one we must have

$$(1.2.29) \qquad y \equiv a \cdot x' + b \pmod N$$

imply that $x \equiv x' \pmod N$. We achieve this by asking that N and the multiplier a be relatively prime.

A perhaps more informative way to proceed is to consider the set of points $\{x_i, x_{i+1}\}$ in the unit square. We divide the unit square into p^2 squares with sides $1/p$ and denote the number of points in the square (ν,μ) by $N_{\nu\mu}$. We want $N_{\nu\mu}/n$ to be close to $1/p^2$. We compute these values and discuss the result, taking n at least equal to 200. We also calculate the covariance and correlation. (Formal statistical tests can be designed, but we must be careful at this point, and note that successive vectors $z_i = (x_i, x_{i+1})$ <u>should</u> be independent.) One may want to modify the algorithm to get better results. We shall find it helpful to start by choosing n small, typing out the sequence and plotting it in the unit square, remembering the form of the algorithm we have used (see Appendix 1).

<u>Assignment</u>. Do the same thing for the sequence of points $\{x_i, x_{i+2}\}$ and discuss your results, either for the FORTRAN subroutine RANDU (ref. 15) or one of the random number generators discussed above.

We can get an analytic complement to this computational study as follows; we shall do it for $x_i = \{ix\}$, x = irrational. If we put $y = \{ix\}$, $z = \{(i+1)x\}$, then if $y \leq 1-x$ we have $z = y+x$ while if $y > 1-x$ we have $z = y+x-1$. This means that the covariance between x and y gets contributions of these two types and since the y are equidistributed, the limit is, for h = 1,

$$(1.2.30) \qquad E(x - \tfrac{1}{2})(y - \tfrac{1}{2}) = R(1) = \lim_{n\to\infty} R^*(1) = \int_0^{1-x} y(y+x)dy + \int_{1-x}^{1} y(y+x-1)dy - \frac{1}{4}$$

$$= \frac{1}{12} - \frac{1}{2}x(1-x)$$

where R(1) and R*(1) are the theoretical and empirical covariances of lag one, respectively (see chapter 4). Since this does not vanish unless $x = \frac{1}{2} \pm 1/(2\sqrt{3})$, we do not get covariance zero. Similarly, we can calculate the covariances for lag $h > 1$ and find that for no value of x do we get all covariances $R(h) = 0$, where $R(h)$ is the covariance of lag h.

There are other iterative schemes for generating pseudo-random numbers with $R(h) = 0$, all $h \neq 0$. Indeed, let $x_i = \{i^2 x\}$; it can be shown that this sequence has the above property. On the other hand, form the second difference of x_i:

$$y_i = \Delta^2 x_i = x_{i+2} - 2x_{i+1} + x_i \tag{1.2.31}$$

It is clear that $\{y_i\} = x =$ constant, so that plotting the triplets $z_i = (x_i, x_{i+1}, x_{i+2})$ in the three-dimensional unit cube yields all the z lying on the parallel surfaces $\{x_{i+2} - 2x_{i+1} + x_i\} = x$; this dependence exists although the covariances vanish. Higher-order polynomials to generate x_i can be dealt with similarly.

This reasoning is more general than it may appear. Indeed, consider an arbitrary rule of generation of the type $x_{i+1} = f(x_i)$ and plot the pairs (x_i, x_{i+1}) in the unit square. If we are not satisfied with $R(1) = 0$, the property of no correlation of lag 1, but want (two-dimensional) equidistribution in the unit square, we must let the graph of f be everywhere dense in the square; a continuous f would not suffice. While this is computationally not very meaningful if interpreted literally, it leads us to the conclusion that <u>a high-quality random number generator of this type must be built on an</u> f <u>with a very complicated graph</u>. The algorithm must be <u>sufficiently complex</u> to generate the sort of pseudo-random behavior we need.

If it were asked that we have equidistribution of dimension 1 and 2, etc., it would be possible to write down generation rules of somewhat different form. One possibility is to take

$$x_i = \{x^i\} \tag{1.2.32}$$

which has this property for almost all $x > 1$. It seems doubtful that such a rule could be used computationally and directly with a particular value of x, but its discrete analogue, the multiplicative congruence method, is used extensively in practice.

<u>Assignments</u>. <u>1</u>. Use some statistical table, say on population size, income data, or some similar real data, to simulate randomness (see ref. 22). Assume that the data are given as decimal numbers with 6 digits. Take out the middle 4 digits and use them to form a decimal fraction between 0 and 1. These numbers will now play the role of $x_1, x_2, \ldots, x_n$. Watch for pitfalls.

Treat them as in earlier sessions at the terminal. Calculate means, variances, covariances, etc. Do they seem to be equidistributed and reasonably independent? You may wish to expose them to some formal statistical test, such as chi-square, Kolmogorov's test or some run test in order to draw conclusions about the randomness of their generation.

If you have a table of numbers that seem fairly random but you would like to make them even "more random", can you suggest some scheme for doing this? You may think of some way of "mixing" the data even more to make them appear more chaotic.

2. Collect a set of physical constants such as the speed of light, the gravitational constant and other measurements (chemical tables may be useful) that are *expressed in some unit* (not necessarily the same unit throughout). Do not use absolute constants such as π and e. In this way you get a set $x_1, x_2, \ldots, x_n$, say, written in decimal form as

$$x = m \cdot 10^y \tag{1.2.33}$$

where y is an integer and the *mantissa* m satisfies $1/10 \leq m < 1$. From your set $\{x_\nu\}$ form the set of mantissas $m_1, m_2, \ldots, m_n$. Does this set appear to have the properties one would expect a random sample from the uniform distribution over (1/10,1) to have? Discuss the result. See eq. (1.2.45).

3. Look up the FORTRAN subroutine RANDU in ref. 15. How is the algorithm constructed? Compare with earlier sections. Form a large sample (it may not be necessary to print the values) and expose it to reasonable tests of randomness. Summarize the results.

This is, of course, a deterministic procedure and we know the rule of generation. We could therefore use this knowledge to *design* a test that would establish the lack of randomness. Suggest such a test; its performance will be discussed in class.

Another rule of generation believed to have good properties is

$$x_{n+1} = M \cdot x_n \pmod{2^{32}} \tag{1.2.34}$$

with

$$M = 477{,}211{,}307 \tag{1.2.35}$$

Look into its properties, as above.

4. We return from the discrete to the continuous. Assume that we have an absolutely continuous probability distribution on the real line with a distribution function F whose density is f. We denote the stochastic variable by X and represent it as a decimal fraction. Move the decimal point p steps to the right and reduce the resulting number modulo 1 to the interval (0,1); call the result X^p. What are the probabilistic properties of X^p?

To explore this we pick a smooth but quite skew probability distribution for X with the frequency function

$$f(x) = 2/(1+x)^3, \quad x > 0 \tag{1.2.36}$$

and hence the distribution function

$$F(x) = \int_0^\infty f(x)\,dx = 1 - 1/(1+x)^2, \quad x > 0 \tag{1.2.37}$$

Let us form the fractional part $Y = \{LX\}$ and derive its probability distribution on the discrete level:

$$q_i = P\{\frac{i-1}{m} \leq Y < \frac{i}{m}\}, \quad i = 1,2,\ldots,m \tag{1.2.38}$$

We get

$$\begin{aligned} q_i &= P\{\frac{i-1}{m} \leq \{LX\} < \frac{i}{m}\} = \sum_{k=0}^{\infty} P\{k + \frac{i-1}{m} \leq LX < k + \frac{i}{m}\} \\ &= \sum_{k=0}^{\infty} [F(\frac{k+[i/m]}{L}) - F(\frac{k+(i-1)/m}{L})] \\ &\cong \sum_{k=0}^{MAX} [(\frac{1}{1 + \frac{k+(i-1)/m}{L}})^2 - (\frac{1}{1 + \frac{k+i/m}{L}})^2] \end{aligned} \tag{1.2.39}$$

The program ALIAS (see Appendix 1, figure 1.5) computes the q's and displays them. Figures 1.6 - 1.8 show the behavior for M = 10 and L = 1, 4, 10. Note that the vertical scale is different in the different graphs.

The graphs show clearly that when L increases the q's tend to become closer to each other: <u>a tendency toward a uniform distribution</u>.

Let us interpret it in terms of the roulette wheel. If X means the total angle the wheel will turn, which will normally be through many revolutions, we can let the classes corresponding to q_i's represent the numbers on the wheel. If there are many classes (L large), we can expect a fairly uniform probability distribution $q_i \approx 1/m$ although f was a fairly arbitrary frequency function. Let us extend this experimental result analytically.

*Since X^p is contained in the interval (0,1) we can do Fourier analysis on it using the set of characters $\{e^{2\pi ikx}; k=0,\pm1,\pm2,\ldots\}$. We therefore form

$$\hat{X}_k^p = E\, e^{2\pi ikX^p} = \int_{-\infty}^{\infty} e^{2\pi ik10^p x} f(x)dx \tag{1.2.40}$$

where we have used the periodic property of the characters. But this expression is the same as $\phi(2\pi k10^p)$ if we introduce the characteristic function, as usual, by

$$\phi(t) = \int_{-\infty}^{\infty} e^{itx} f(x)dx \tag{1.2.41}$$

Appealing to the Riemann-Lebesgue lemma (ref. 14, p. 469), we see that

$$\lim_{p\to\infty} \hat{X}_k^p = 0 \tag{1.2.42}$$

for $k \neq 0$; when $k = 0$ the limit is, of course, 1. The distributions tend to the uniform distribution, which proves the

Theorem. If X has an absolutely continuous distribution, then $\{10^p X\}$ converges in distribution to R(0,1) as $p \to \infty$.

Note that the original form of the distribution does not matter as long as it is absolutely continuous.

The practical interpretation of this result requires some care, since the stochastic variables with which we deal are always given with a finite number of digits. We must, therefore, not take p too large, but we can hope that a moderate value of p will result in an approximately uniform distribution. This was studied numerically on page 15.

Let us now analyze the second problem on page 15. We can write the mantissa m(x) of x as

$$m(x) = 10^{\{\log_{10}x\}-1} \tag{1.2.43}$$

Let us assume that we have drawn our sample from a population of universal constants where a probability distribution of m is given. Note that this innocuous-looking assumption - that such a distribution exists - is not necessarily satisfied a priori. (One of the several arguments which can be raised against existence is that the population of universal constants is not defined.) Let us assume, as may seem natural, that the distribution should not depend on the units of measurement chosen. Then a change of scale should not result in any change of distribution, so that $\{\log_{10}x\}$ and $\{\log_{10}cx\} = \{\log_{10}x + \log_{10}c\}$ should have the same distribution. We now again use Fourier analysis and find that $\{\log_{10}x\}$ must have a uniform distribution over (0,1). *Indeed, if the Fourier coefficients with respect to the distribution are ϕ_k, k = 0,±1,±2,..., then the Fourier coefficients of the distribution corresponding to the expression $\{\log_{10}x + \log_{10}c\}$ are, for any integer k,

(1.2.44) $$\phi_k \cdot \exp(2\pi ik \log_{10}c)$$

which should be equal to ϕ_k for all k. But this implies that $\phi_k = 0$ except for k = 0, which means that the distribution (of $\log_{10}x$) must be uniform, as asserted.* Hence the mantissa m must be distributed according to the rule

(1.2.45) $$P(m\leq u) = P(10^{y-1}\leq u) = P(y\leq 1+\log_{10}u) = 1 + \log_{10}u, \quad 1/10 \leq u < 1$$

and frequency function = $1/u \ln 10$. This reasoning is not quite watertight (why not?) but it throws some light on the departure from uniform mantissa distribution that has repeatedly been observed in practice for numbers generated in this particular fashion.

Assume that we have been able to produce, in one way or another, a sequence $\{x_n\}$ of binary variables, and that the sequence behaves reasonably well as far as independence is concerned. There is, however, a certain element of bias: the value $P(x_n=1) = p$ deviates too much from 1/2, so that $\varepsilon = p - 1/2$ is not negligible. We can then do better by introducing the new sequence {y} where

(1.2.46) $$y_n = \begin{cases} 1 & \text{if } x_{2n} = x_{2n-1} \\ 0 & \text{otherwise} \end{cases}$$

Indeed,

(1.2.47) $$P(y_n=1) = p^2 + (1-p)^2 = \frac{1}{2} + 2\varepsilon^2$$

which is closer to the ideal value 1/2 than p. Note, however, that in this way we get a new sequence which is only half as long as the original one. The procedure can be iterated, now starting, instead, from $\{y_n\}$, and we get as close to 1/2 as we wish but at the cost of getting shorter and shorter sequences.

Another way of reducing bias is the following, which has been applied a great deal. Group a sequence {X} by twos: $(x_1,x_2),(x_3,x_4),(x_5,x_6),\ldots$. Throw away all pairs of the form (0,0) and (1,1). In the remaining sequence of pairs replace an appearance of (0,1) by 0 and of (1,0) by 1. Call the derived sequence {y}. Then

(1.2.48) $$P(y=0) = P(x_i=0,x_{i+1}=1 \mid \text{given that either } x_i=0,x_{i+1}=1 \text{ or } x_i=1,x_{i+1}=0)$$

$$= \frac{(1-p)p}{(1-p)p + p(1-p)} = \frac{1}{2}$$

so that the sequence {y} has no bias in its (marginal) distribution. Note that this reasoning relies on the x_n being independent of each other.

Related ideas have been used to reduce dependence between successive numbers when such a dependence is suspected. Simply by picking out a sub-sequence, say of every r^{th} number, one may sometimes reduce the dependence. If the original sequence is from a Markov chain with transition probability matrix P, then the derived sequence will instead have the transition probability matrix P^r. Assume that the original sequence has been generated by some physical device and that one suspects the existence of some weak dependence. P^r will be better if, say, the eigenvalues of P are less than one in absolute value except for the single eigenvalue 1. (Discuss this in class and ask for other suggestions in this direction.)

Assignment. Let $x_1,x_2,x_3,\ldots$ be a Markov chain with the four-state probability transition

(1.2.49) $$P = \begin{bmatrix} .80 & .1 & .05 & .05 \\ .2 & .7 & .05 & .05 \\ .1 & .6 & .1 & .1 \\ .05 & .05 & .1 & .5 \end{bmatrix}$$

Calculate empirical estimates of the covariances

$$C_h = E(x_n - m)(x_{n+h} - m)$$

from a simulation experiment, assuming that we start from the equilibrium distribution $p = (p_1, p_2, p_3, p_4)$ satisfying

$$pP = p \tag{1.2.50}$$

If we take only every r^{th} value of x, how does the covariance (of lag 1 for the new sequence) behave?

The following direct solution of (1.2.50) has been proposed by Mr. John F. Pieper, a student at Brown University.

When (P-I) is non-singular, (1.2.50) can be reduced to

$$p = (P-I)^{-1}.0 = 0$$

Thus, the only cases of interest are those where (P-I) is non-singular. Eq. (1.2.50) is normally solved by iteration rather than by direct solution of

$$pA = 0 \quad (A \equiv P-I) \tag{1.2.50a}$$

That is, p is found to be $\lim_{n\to\infty} p_n$, where $p_n = p_{n-1}P$, p_0 = initial guess. The following discussion shows how one can solve (1.2.50a) directly for the vector p, obtaining an exact answer with less effort than would normally be used for an approximation.

Consider the matrix A; if it were non-singular, we could reduce it, via row or column reduction methods, to the identity matrix. Here this is not possible as A is singular; however, row or column reduction methods applied to A would, in effect, isolate and expose the singularity(ies) inherent in A.

Let $\{E_1, E_2, \ldots, E_n, \ldots\}$ be matrices corresponding to elementary operations (multiply a row (column) by a scalar; add a row (column) to another; interchange two rows (columns)). Multiplying a matrix from the left by one of these E_i matrices performs a row reduction and from the right a column reduction operation. Thus, were we to form the matrix Q = the column reduction of A, we would form $Q = A \cdot E_1 \cdot E_2 \cdot \ldots \cdot E_n \cdot \ldots$.

Now, all elementary matrices are non-singular; the product of two non-singular matrices is also non-singular; thus $E \equiv E_1 \cdot E_2 \cdot \ldots \cdot E_n \cdot \ldots$ has an inverse. Then we may write $A = Q \cdot E^{-1}$, where Q is singular and represents the furthest possible reduction of A. Note that we have, in effect, separated the singular matrix A into two portions: one, Q, containing the basic singularity(ies) of A, and one, E^{-1}, representing non-singularities that serve only to hide the basic nature of A. As all the useful information about A is contained in Q, we throw E^{-1} away:

$$pA = 0 \qquad pQE^{-1} = 0 \qquad \text{or} \qquad pQ = 0$$

Due to the simple form of Q, this can readily be solved.

Column reduction of A will eventually result in a column of all-zero entries; by suitable manipulation, this should be made the last column. A "double" singularity in A will result in two all-zero columns in Q, etc. In most cases Q will be of the form

$$\begin{bmatrix} 1 & 0 & 0 & \cdots & 0 & 0 \\ 0 & 1 & 0 & & 0 & 0 \\ & \cdots & \cdots & \cdots & \cdots & \\ 0 & 0 & 0 & & 1 & 0 \\ -\lambda_1 & -\lambda_2 & -\lambda_3 & \cdots & -\lambda_{n-1} & 0 \end{bmatrix}$$

in which case the answer is

$$\begin{aligned} p_1 &= \lambda_1 p_n \\ p_2 &= \lambda_2 p_n \\ &\vdots \\ p_{n-1} &= \lambda_{n-1} p_n \\ p_n &= p_n \end{aligned}$$

Then, if the p_i represent probabilities such that $\sum_{i=1}^{n} p_i = 1$, or if some other condition on the p_i is known, an exact solution can be achieved.

Occasionally Q will be of such a form that an infinite number of solutions exists; for example,

$$\begin{bmatrix} 1 & 0 & 0 \\ -1 & 0 & 0 \\ 0 & 0 & 1 \end{bmatrix}$$

or

$$\left.\begin{array}{l} p_1 = p_2 \\ p_3 = p_3 \end{array}\right\} = \text{solution}$$

Comparative Example. Let P be a matrix of transition probabilities and p be the equilibrium distribution vector associated with a given state system.

$$\sum_{j=1}^{3} P_{ij} = 1 \quad \text{for all } i \; ; \qquad \sum_{i=1}^{3} p_i = 1$$

Let

$$P = \begin{bmatrix} .5 & .2 & .3 \\ .2 & .2 & .6 \\ .4 & .2 & .4 \end{bmatrix}$$

By iteration:

$$\begin{array}{llll} p_0 = & 1 & 0 & 0 \\ p_1 = & .5 & .2 & .3 \\ p_2 = & .41 & .2 & .39 \\ p_3 = & .401 & .2 & .399 \end{array} \qquad \text{the } \frac{n+1}{2} \text{ iteration}$$

- - - - - - - - - - - - -

$$p = [.4 \quad .2 \quad .4]$$

Direct solution:

$$A = P - I = \begin{bmatrix} -.5 & .2 & .3 \\ .2 & -.8 & .6 \\ .4 & .2 & -.6 \end{bmatrix} \qquad Q = \begin{bmatrix} 1 & 0 & 0 \\ 0 & 1 & 0 \\ -1. & -.5 & 0 \end{bmatrix}$$

$$\begin{array}{l} p_1 = p_3 \\ p_2 = \frac{1}{2} p_3 \\ \underline{p_3 = p_3} \\ \Sigma p_i = 2.5 p_3 = 1 \end{array} \qquad \text{so } p = [.4 \quad .2 \quad .4]$$

Processing time required: if P is of size n×n, then

i) for each level of iteration we require n^2 multiplications and (n^2-n) additions;

ii) for completely column reducing an n×n matrix we reguire $(n^3+n^2-2n)/2$ multiplications and $(n^3-n)/2$ additions and n divisions;

iii) for $(n+1)/2$ levels of iteration, the exact solution requires n fewer multiplications, the same number of additions and n additional divisions compared to iteration;

iv) for greater than $(n+1)/2$ levels of iteration, the exact solution is faster as far as processing is concerned. Also, it gives the exact answer and is independent of an initial estimate of p.

*To a reader familiar with measure theory, the following remarks may help clarify the situation. The computational approach is helpful when comparing our intuitive idea of randomness with the mathematical model we use. The latter, expressed in the language of measure theory, does not tell the whole story of how these two notions, the intuitive and the formal, are related.

To be concrete, let us think of a sequence of trials in each of which we throw an unbiased coin and record the outcome $x = (x_1, x_2, x_3, \ldots)$ when $x_i = 1$ for heads and 0 otherwise. The formalization of this consists in introducing a probability measure P in the sample space $X = B^{\infty}$, where $B = \{0,1\}$. The measure is given as a product of marginal measures assigning the probability 1/2 to each of the values 0 and 1. It is well known how the classical results from probability theory can be expressed in terms of P. The strong law of large numbers, to mention just one of many results, says that

$$\lim_{n\to\infty} \frac{x_1+x_2+\ldots+x_n}{n} = 1/2 \tag{1.2.51}$$

almost certainly. In other words, the limit exists and is equal to one-half except for a subset $N \subset X$ which is small in the sense that $P(N) = 0$. Here X is the set of all (infinite) x-sequences.*

To get closer to the behavior of the individual sequences, one could try to define a particular sequence x as random if it has the limit indicated above. This does not really make much sense, since we would then have to accept a sequence like (0,1,0,1,0,1,...) as random. Its behavior is too simple. We would have to supplement the definition by some selection invariance requirement: for instance, we should require to get the same limit for sub-sequences. We must now rule out the possibility that we select certain sub-sequences with no apparent randomness, for example the sub-sequences consisting of all 1's. In the early development of this approach, the theory of "Kollektive" of von Mises, this was sometimes expressed by saying that the selection should be made without looking at the values of the components of x. Since x is just any fixed binary sequence, it is not quite clear exactly what this restriction should mean.

Let us look at the notion of randomness from the point of view of a computer programmer. He wants to write a program for his machine to generate a binary sequence exhibiting such properties as we would expect of a random sequence. We turn the problem around, in that he is given a fixed and finite sequence and asks for the shortest program that will generate x; call the length of the program $\lambda(x)$. This length will depend on the machine he uses. On the other hand, with two machines M_1 and M_2, if x is a long vector, we can expect that the two values of $\lambda(x)$ will be asymptotically close to each other; the instructions needed for a translator from M_1 to M_2 are fixed and will be asymptotically negligible. The complexity $\lambda(x)$ is therefore asymptotically machine-independent when the length of the sequence increases.

A suggestion has been made by Kolmogorov and explored by Per Martin-Löf (ref. 17) that one introduce the notion of randomness via the complexity $\lambda(x)$. A sequence will be called random if its complexity is close to the largest possible value. To make this precise, we replace the real computer by a Turing machine and the formal definitions will be introduced in the spirit of recursive function theory. It can be shown not only that this is possible in a logically consistent way, but that sequences accepted as random in this approach possess all the standard limit properties of classical probability theory such as the law of large numbers, the law of the iterated logarithm and the central limit theory.

If one accepts this way of introducing randomness, it is clear that the usual deterministic generators of random numbers do not seem very attractive; they are almost antithetic to random sequences. Indeed, the usual multiplicative-additive congruence generators can be programmed with just a few FORTRAN statements. And not only that; it is practically impossible to generate randomness by arithmetic means since, by definition, the program must be very long to lead to random sequences.

It is too early to try to judge this approach and it is not yet clear what practical consequences it will have. Almost all practitioners of the Monte Carlo method and artificial sampling employ arithmetic random number generators, and it may appear questionable, in the light of what has been said above, that these really simulate randomness. We shall examine the basic problem of the rationale of the Monte Carlo technique in another section. There we shall see that it is possible to motivate this technique, but in a different way from the usual one.

We now turn to generators of pseudo-random sequences used in practice. For computational economy it is desirable that the sequence be computed by a recurrence relation

$$x_{n+1} = f(x_n, x_{n-1}, x_{n-2}, \ldots, x_{n-r}) \tag{1.2.52}$$

where f is a function that is easy to compute on the available machine. In particular, linear functions come to mind, and also the fact that the reduction modulo m is easily carried out if m is chosen as the largest integer representable in the machine, plus one. Taking the order r as low as possible, $r = 0$, (1.2.52) then reduces to

$$x_{n+1} \equiv \lambda x_n + \mu (\text{mod } m) \tag{1.2.53}$$

(where λ and μ are integers), which specifies the pseudo-random sequence $\{x_n\}$ if we decide on the starting value x_0. In a binary machine it is natural to choose $m = 2^w$, where w is the word length. We can achieve some computational simplicity by asking that λ be of the form 2^k+1, since this corresponds simply to one shift and one add operation.

It is important to choose the parameters λ, μ in such a way that $\{x_n\}$ exhibits some of the features of randomness we want. Any congruence method of order r must obviously be periodic: looking at the vectors $\xi_n = (x_n, x_{n-1}, \ldots, x_{n-r})$ taking values in a discrete hypercube of dimension $r+1$, we see that this cube contains m^{p+1} distinct points, implying that the period cannot be greater than m^{p+1}.

The congruence generator (1.2.53) is called mixed if both λ and μ are different from zero. If $\mu = 0$ we speak of a multiplicative generator, and if $\lambda = 0$ we call it additive (see the discussion above). The maximal period is m. If the maximal period is attained it follows that the x_n's run through all the residue classes exactly once during a period. We need to know when the maximum period is attained and, when it is not, how large a period we get and what residue classes are generated. The latter question is related to the degree of the (discrete) one-dimensional equidistribution which has been generated.

It should be pointed out here that answers to these questions are not sufficient to enable us to judge the performance of the generator. In addition we must know how good higher-dimensional equidistributions we get and how much dependence we have in the sequence. Dependence, as well as other similar notions, is meant in the sense of asymptotic distribution: because of the periodic nature of the sequences the limits exist.

This is not the proper place to go into a detailed discussion of the periods and distribution properties of the congruence generators. Relevant papers for information on this topic are those by Jansson (ref. 1) and Coveyou and MacPherson (ref. 2).

We give, in figure 1.9 of Appendix 1, a program to generate the bivariate distribution of the multiplicative congruence generator mod N, with multiplier N. Illustrations of graphic output from this program are given in figures 1.10, 1.11 and 1.12. In figure 1.10, M (= 3) and N (= 20) are relatively prime and the results are quite good. In figure 1.11, the value of M (= 19) is obviously too large. The multiplier M should neither be very small nor very large compared to the module N, and in addition M and N should be relatively prime. Figure 1.12 (M = 7, N = 20) also shows reasonably good results.

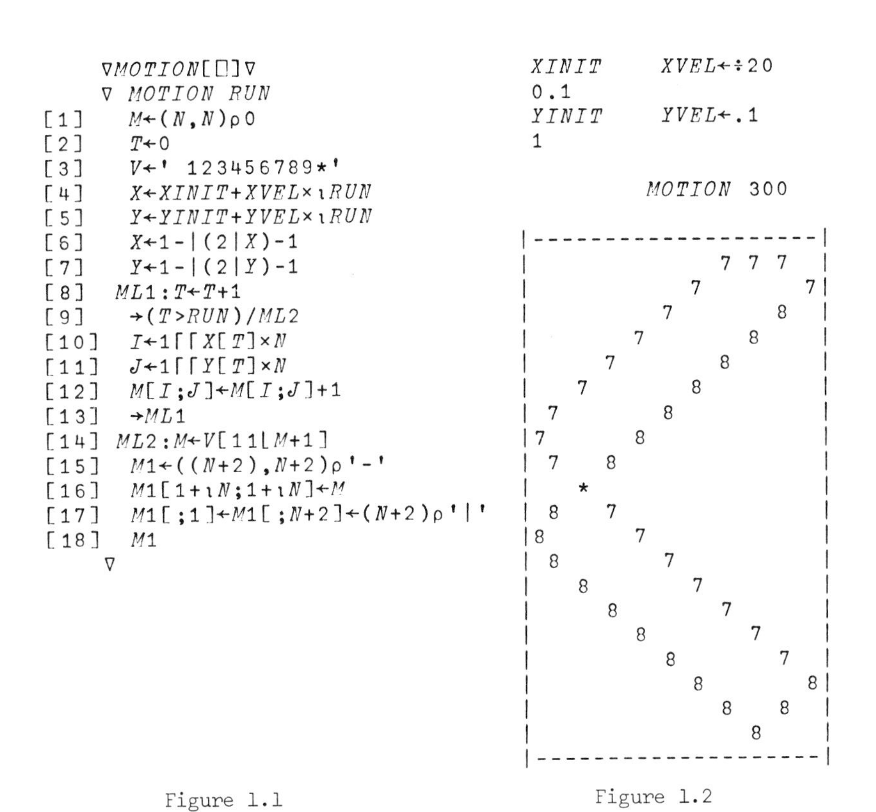

Figure 1.1

Figure 1.2

The above program MOTION generates S_{ij} of eq. (1.2.27).

The parameters XINIT and YINIT are the coordinates of the initial position of the point at time T = 0. RUN is the number of time-points during the run.

Figure 1.2 gives the results of MOTION for the stated parameters XVEL = 1/20, RUN = 300. The numerals indicate the number of times the mass point passes through a certain point within the square.

```
 XINIT    XVEL←÷30
0.1
 YINIT    YVEL←.1
1
 MOTION 300

|--------------------|
|           5 5 5    |
|       5 5       5 5|
|     5           5  |
| 5 5         5 5    |
|5          5        |
| 5 5   5 5          |
|     *              |
| 5 5   5 5          |
|5          5        |
| 5 5         5 5    |
|     5           5  |
|       5 5       5 5|
|           5   5    |
|           5 * 5    |
|         5       5  |
|     5 5         5 5|
|   5           5    |
|55         5 5      |
| 5       5          |
|   5 5 5            |
|--------------------|
```

Figure 1.3

```
 XINIT    XVEL←.11
0.1
 YINIT    YVEL←.1
1
 MOTION 300
|--------------------|
|2  3   3 3   3   2  |
| 3   3     3   3   2|
| 4   3   3 3   2    |
|3  3   3     3   3  |
|   4   3 3   2   3  |
| 3   3     3   3   3|
| 3   4   3 2   3    |
|3  3   3     3   3  |
|   3   4 2   3   3  |
| 3   3     3   3   3|
| 3   3   4 3   3    |
|3  3   2     3   3  |
|   3   3 3   3   3  |
| 3   2     4   3   3|
| 3   3   3 3   3    |
|3  2   3     4   3  |
|   3   3 3   3   3  |
| 2   3     3   4   4|
| 3   3   3 3   3    |
|1  3   3     3   4  |
|--------------------|
```

Figure 1.4

```
∇ ALIAS[⎕]∇
    ∇ M ALIAS L
[1]    Q←Mρ0
[2]    K←0
[3]   AL1:Q←Q+÷((1+(÷L)×K+(÷M)×¯1+⍳M))*2
[4]    Q←Q-÷(1+(÷L)×K+(÷M)×⍳M)*2
[5]    K←K+1
[6]    →(K≤MAX)/AL1
[7]   AL2:30 PLOT Q
    ∇
```

Figure 1.5

The program ALIAS computes eq. (1.2.39).

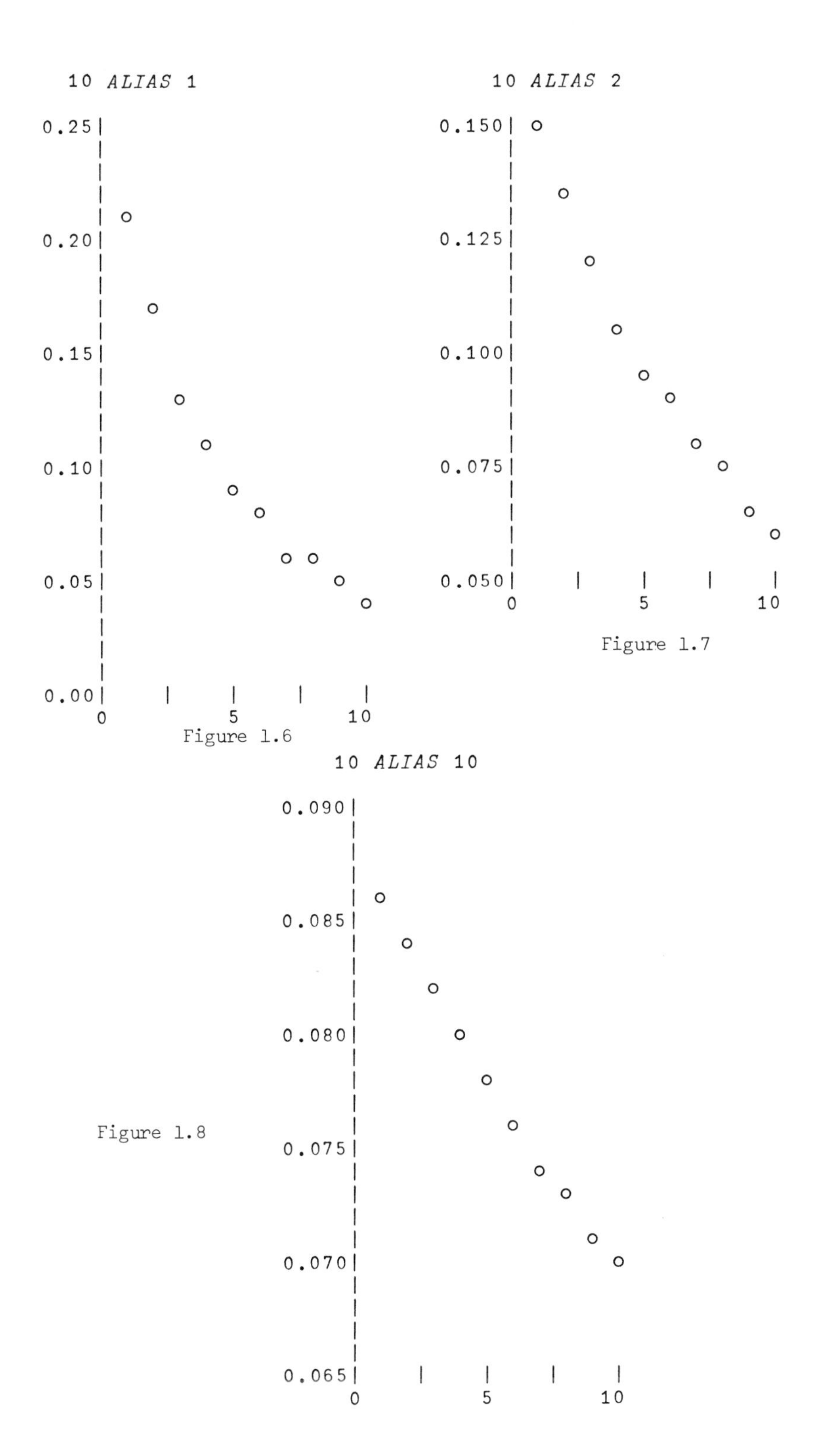

Figure 1.6

Figure 1.7

Figure 1.8

```
      ∇BIVARIATE[⎕]∇
    ∇ M BIVARIATE N
[1]   ('.×')[1+(¯1+⍳N)∘.=N|M×¯1+⍳N]
    ∇
```

Figure 1.9

```
      3 BIVARIATE 20
×...................
.......×............
..............×.....
.×..................
........×...........
...............×....
..×.................
.........×..........
................×...
...×................
..........×.........
.................×..
....×...............
...........×........
..................×.
.....×..............
............×.......
...................×
......×.............
.............×......
```

Figure 1.10

```
      19 BIVARIATE 20
×...................
...................×
..................×.
.................×..
................×...
...............×....
..............×.....
.............×......
............×.......
...........×........
..........×.........
.........×..........
........×...........
.......×............
......×.............
.....×..............
....×...............
...×................
..×.................
.×..................
```

Figure 1.11

```
      7 BIVARIATE 30
×.............................
.............×................
..........................×...
.........×....................
......................×.......
.....×........................
..................×...........
.×............................
..............×...............
...........................×..
..........×...................
.......................×......
......×.......................
...................×..........
..×...........................
...............×..............
............................×.
...........×..................
........................×.....
.......×......................
....................×.........
...×..........................
................×.............
.............................×
............×.................
.........................×....
........×.....................
.....................×........
....×.........................
.................×............
```

Figure 1.12

CHAPTER 2:

SIMULATION

2.1 Simple Monte Carlo.

In this chapter we shall look into one particular computational technique that is being used (and misused) extensively in analyzing stochastic systems: the method of Monte Carlo. The material is closely related to that presented in chapter 1, but is of a more technical nature. In later chapters the method will be applied to many special cases of practical interest, but here the attitude is methodological, with some emphasis on computational efficiency.

Most of the quantities we compute in mathematical statistics are either integrals (mean values, moments, characteristic functions) or derived from integrals (central moments, quantiles). It is very often not possible to evaluate the integral in closed form and it may be difficult or at least uneconomical to appeal to the standard methods of numerical quadrature. We can then try a Monte Carlo approach.

The idea of Monte Carlo is simple enough. Say that we want to find the value of the integral

$$I = \int_X f(x)\,dx \tag{2.1.1}$$

where dx denotes a bounded measure over the space X and f is real and integrable with respect to dx. The domain of integration X could be an interval on the real line and dx some probability distribution on it, or we could consider more complicated spaces. Generate a sample $x_1, x_2, \dots, x_n$ of values from X drawn according to the probability measure $\frac{1}{M}dx$ where M is the total dx-measure of X; assume, to simplify notation, that M = 1. It then follows from the law of large numbers that

$$I_n^* = \frac{1}{n}\sum^{n} f(x_i) \tag{2.1.2}$$

converges almost certainly to I as n tends to infinity. If f is quadratically integrable, so that the formal "variance" $\sigma^2(f)$, namely

$$\sigma^2(f) = \int [f(x) - I]^2\,dx, \tag{2.1.3}$$

exists, we know that I_n^* has standard deviation $(1/\sqrt{n})\,\sigma(f)$. If n is large, we can state that the value I of our integral is covered by the interval $(I_n^*-\lambda[\sigma(f)/\sqrt{n}],$ $I_n^*+\lambda[\sigma(f)/\sqrt{n}]$ with probability approximately equal to $2\Phi(\lambda)-1$, where Φ is the distribution function of the normed, normal distribution $\Phi = N(0,1)$. If we do not know $\sigma(f)$, and we seldom do, we will have to use the sample estimate of it.

We get in this manner an approximation for the value of the integral as well as a confidence interval around this value. This is one version of the Monte Carlo method; it is completely straightforward and we shall see in section 2.3 that we can often do a great deal better than this.

This example does not explain, however, why we use simulation rather than one of the many classical methods of numerical quadrature. To bring this out more clearly, we turn to higher-dimensional situations where the power of simulation becomes clearer.

Let us look, for illustration, at the following problem. We have a continuous probability distribution F in the (x,y) plane and we want to test independence between the x and y coordinates. To do this we collect a sample $(x_1,y_1),(x_2,y_2),\ldots,$ (x_N,y_N) and form the empirical medians M_x^* and M_y^* (assume that N is odd). Introduce the numbers

$$\text{(2.1.3a)} \qquad \begin{aligned} N_1 &= \text{number of points in the first quadrant; } x > M_x^*, y > M_y^* \\ N_3 &= \text{number of points in the third quadrant; } x < M_x^*, y < M_y^* \end{aligned}$$

and form the statistic

$$\text{(2.1.3b)} \qquad t = \frac{N_1+N_3}{N}$$

In order to construct a test of independence built on t we need the probability distribution of t, at least approximately, for various F.

This is, however, equivalent to evaluating certain integrals in R^{2N}. If we had access to a closed-form solution, and this were amenable to computing, we would use it; otherwise we have to choose between two alternative actions.

The first is to use a large sample approximation, which is simple in principle although possibly tedious. The trouble is that we may need the result for such a small sample size N that we do not trust an approximation which assumes the sample size to be large.

The other way out is to use Monte Carlo to give us an idea of the approximate distribution of t for various distributions F. We take n samples of length N from the population F and form the empirical distribution function for the n values of t computed from the samples obtained. We use this distribution as an approximation to the true distribution of t.

The Monte Carlo method is used very frequently, often in an uncritical and inefficient way. It has some obvious advantages:

a) It requires no regularity assumptions on f except square integrability. (Compare it with the traditional error estimates of numerical integration in terms of derivatives or other smoothness requirements!) This may be of help since f is often a very complicated function whose regularity properties are difficult to find.

b) It is feasible even in high-dimensional spaces for which numerical integration fails, say dimension ≥ 10. In the present context this is relevant since we can very well run into 100-dimensional problems or larger in computational probability studies.

c) It is easy to apply, requires little or no analysis of the problem (but this is not true of the better designs we shall look into later).

It has, however, some disadvantages also:

a) The error bound is not a certain one but involves some randomness; this is more nearly a psychological difficulty than a real one.

b) The statistical precision is often very low; more about this later.

c) We need random numbers and we already know that the generation of reasonably "random" sequences is not trivial (in section 2.2 we shall indicate that this observation may be less relevant than one would think at first sight).

We shall below use Monte Carlo only to study problems presented in a stochastic form. It should be mentioned, however, that the method can be applicable in situations where no randomness appears in the original presentation of the problem. We shall not deal with such cases in what follows.

Before going ahead to a more detailed discussion of some Monte Carlo refinements, let us point out the following. Say that we have decided upon a way of generating pseudo-random numbers simulating the population R(0,1). It is then an easy matter to get other distributions, for instance one with some given distribution function F. Indeed, if F is discrete, with the probabilities $P_1, P_2, \ldots$ associated with the values

$x_1, x_2, \ldots$, then we put

$$(2.1.4) \qquad z = x_i \quad \text{if} \quad \sum_1^{i-1} P_j \leq x < \sum_1^{i} P_j$$

where x is the pseudo-random number from R(0,1). It is clear that z will be distributed according to F. This can be directly modified for the case when F is continuous. We put

$$(2.1.5) \qquad z = F^{-1}(x)$$

where F^{-1} is the inverse function of F. In both cases z will have the distribution function F.

As an example, assume that we want to generate exponentially distributed numbers such as, say, in simulation of a Markovian queueing problem. For $F(x) = 1-e^{-x}$ we get $z = -\log(1-x)$, which has the same distribution as $z' = -\log x$, $x = R(0,1)$. By adding k such z's we can simulate the Erlang distribution with frequency function

$$(2.1.6) \qquad \frac{1}{k!} x^k e^{-x}, \qquad x > 0$$

This procedure is simple enough and if only a moderate-size Monte Carlo experiment is planned it will be acceptable. If many random numbers are needed we may want higher computational efficiency by applying some special trick for simulating just F. To give an idea of what can be achieved in this way, let us say that we have for some natural number r

$$(2.1.7) \qquad F(x) = 1 - (1-x)^r, \quad 0 \leq x \leq 1$$

Then, rather than inverting F, it may (depending upon the properties of the computing facilities to which we have access) be more economical to generate r values of x from R(0,1) and define

$$(2.1.8) \qquad z = \min(x_1, x_2, \ldots, x_r)$$

Direct calculation shows that z obeys the law in (2.1.7). An example in a slightly different spirit is to generate normally distributed pseudo-random numbers starting from a pair (u,v) independently drawn from R(0,1). Putting

$$(2.1.9)\qquad \begin{aligned} z_1 &= R \cos \phi \\ z_2 &= R \sin \phi \end{aligned}$$

with

$$(2.1.10)\qquad \begin{aligned} \phi &= 2\pi u \\ R &= \sqrt{-2 \log v} \end{aligned}$$

makes z_1 and z_2 independent and $N(0,1)$, as can be shown by a straightforward argument. Many variations on this theme are possible.

The most common method of simulating the normal distribution is probably via the FORTRAN subroutine GAUSS with

$$(2.1.11)\qquad Z = X_1 + X_2 + \ldots + X_{12} - 6.0$$

Here the X's are independent $R(0,1)$, for example from RANDU, and we put our trust in the central limit theorem with sample size 12: $Z \simeq N(0,1)$.

Multivariate distributions are in principle not difficult to handle. Assume that we want to generate a stochastic vector $z = (z_1,z_2,\ldots,z_n)$ with normal distribution and the parameters

$$(2.1.12)\qquad \begin{aligned} Ez_i &= m_i; & i &= 1,2,\ldots,n \\ \operatorname{cov}(z_i,z_j) &= m_{ij}; & i,j &= 1,2,\ldots,n \end{aligned}$$

Since the covariance matrix

$$(2.1.13)\qquad \{M = m_{ij};\quad i,j = 1,2,\ldots,n\}$$

is non-negative definite we can form the square root (or rather one of the square roots) $S = M^{1/2}$. Start from the stochastic vector $x = (x_1,x_2,\ldots,x_n)$ with components normally distributed, means zero, standard deviations one and all independent. Put

$$(2.1.14)\qquad Z' = m' + Sx'$$

so that $Ez' = m'$ = column vector with entries m_i and

$$(2.1.15)\qquad \text{covariance matrix} = E(z-m)'(z-m) = SEx'xS' = SS' = R$$

as desired.

Simulation of stochastic processes is not much more difficult if their structure is not too complex. Indeed, let us generate a Gaussian stationary stochastic process x_t, $t = 1,2,\ldots,n$, with mean value function identically zero and covariance function r_t corresponding to a spectral density $f(\lambda)$. We could apply the procedure in (2.1.14), but if n is large there are computationally better ways of doing this, avoiding operations on the large matrices appearing in (2.1.13) - (2.1.15). Approximate 1/f by a non-negative trigonometric polynomial; according to a classical theorem of Fejér such a polynomial can be written as the absolute square $|P|^2$ of some other trigonometric polynomial:

$$P = a_0 e^{ip\lambda} + a_1 e^{i(p-1)\lambda} + \ldots + a_p \tag{2.1.16}$$

(Fejér's theorem reads: any non-negative trigonometric polynomial, say in complex form, can be written as the squared absolute value of a trigonometric polynomial.) But the spectral density $1/|P|^2$ is that of the autoregressive stochastic difference equation

$$a_0 x_{t+p} + a_1 x_{t+p-1} + \ldots + a_p x_t = \varepsilon_t \tag{2.1.17}$$

where ε_t is white, Gaussian noise.

We can use (2.1.17) in two ways. We can either start with p-1 arbitrary values for $x_{N+p-1}, x_{N+p-2}, \ldots, x_N$ (N some negative integer) and solve (2.1.17) successively from $t = N$ on after generating the ε-sequence in the standard manner. If N is large enough the process should be close to its equilibrium distribution for $t = 1$; we discard, of course, the values $x_N, x_{N+1}, \ldots, x_0$. The speed of convergence is geometric and is good if the solutions of the characteristic equation

$$a_0 z^p + a_1 z^{p-1} + \ldots + a_p = 0 \tag{2.1.18}$$

are sufficiently different from 1 in absolute value (be careful with what "square root" P you form 1/f! This is related to the location of the zeroes of (2.1.18) and should be discussed in class.)

The other method consists of first generating a sample $x_1, x_2, \ldots, x_{p-1}$ with the right equilibrium distribution by the method of (2.1.14) and for $t \geq p$ solving the stochastic difference equation as before. This is safer but more time-consuming.

Markov processes, with or without linear structure, can be generated computationally by just simulating the behavior inherent in the definition of the transition kernel. If this is done by solving a stochastic differential equation some thought must be given to the question of what numerical procedure to use; more about this later.

2.2 Assignments.

1. Consider the integral

$$I = \int_0^1 e^{-1/x}\, dx; \tag{2.2.1}$$

it is a value of an incomplete Γ-function. Determine the value of I by three Monte Carlo experiments with samples of size n = 10, 50, 100 respectively. Replicate this procedure 10 times and give an estimate for I as well as a confidence interval. Use one of the multiplicative congruence methods for generating the random numbers. Look up the value of I in a table of the incomplete Γ-function.

Do the same but using the sequence $x_i = \{i/\sqrt{2}\}$ of pseudo-random numbers generated by an additive congruence rule.

Discuss the results and the accuracy you get and compare with numerical quadrature.

2. Simulate a three-state Markov chain with a transition probability matrix

$$M = \begin{bmatrix} 1/4 & 1/2 & 1/4 \\ 1/3 & 1/9 & 5/9 \\ 1/12 & 1/12 & 5/6 \end{bmatrix} \tag{2.2.2}$$

Start the chain in the state 1 and get sequences of length 3, 5, 15 and sample size n = 100. Use the result to study the convergence to the equilibrium distribution. How is the speed of convergence related to the eigenvalues of M? Find them.

*2.3 Randomness and Monte Carlo.

How serious is it that we do not use truly random sequences in a Monte Carlo experiment, but just the pseudo-random numbers?

Let us look at a special case. Say that we use the additive congruence method with $x_i = \{i\alpha\}$, where α is an irrational number in (0,1) (discuss the results of (2.2) at this point). Say also that, just as an example of no practical interest, we want to integrate the function x - 1/2 over (0,1); we form the sums

$$S_N = \sum_{1}^{n} [\{i\alpha\} - 1/2] \tag{2.3.1}$$

We already know (see chapter 1) that

$$S_N = N \int_0^1 (x - 1/2)dx + o(N) = o(N) \tag{2.3.2}$$

We want to get a more precise bound for S_N/N than just o(1). To succeed with this we must use better analytical tools than before; we need some <u>diophantine analysis</u> to study how well irrational numbers can be approximated by rationals.

First, let us remind ourselves of some elementary properties of continued fractions; we write the irrational number α as a continued fraction:

$$\alpha = a_0 + \cfrac{1}{a_1 + \cfrac{1}{a_2 + \cfrac{1}{a_3 + \cdots}}} = [a_0, a_1, a_2, \ldots] \tag{2.3.3}$$

The truncated continued fractions are denoted by

$$\frac{p_n}{q_n} = [a_0, a_1, \ldots, a_n] \tag{2.3.4}$$

with relatively prime numerator and denominator. Let us note some elementary properties of continued fractions and then develop the tool that is needed. If α is irrational the continued fraction (2.3.3) does not terminate.

The denominators q_i are positive integers and form an increasing sequence.

The sequence of <u>principal convergents</u> p_n/q_n is strictly increasing for even n and strictly decreasing for odd n with the same limiting value α.

The best rational approximations to α are the principal convergents (a fraction p/q is called a best approximation to α if:

$$|q'\alpha - p'| > |q\alpha - p| \tag{2.3.5}$$

for any $1 \leq q' < p'$). We also have $1/(2q_{n+1}) < |q_n - p_n| < 1/(q_{n+1})$.

One of the basic problems in diophantine analysis is to find how small we can make $|q\alpha - p|$ if $q \leq Q$, where Q is some given natural number. It is clear that we can make it smaller than $1/Q$. Indeed, split $(0,1)$ into Q subintervals of equal length. If we introduce the $Q+1$ numbers $\{n\alpha\}$, $n = 0,1,2,\ldots,Q$, it is clear that two of them must fall in the same subinterval, so that $\{\nu\alpha\}$ and $\{\mu\alpha\}$ are in one of them, say in $(\frac{k}{Q},\frac{k+1}{Q})$. But this implies that $|\lambda\alpha - i| < 1/Q$ for some integer i, where $\lambda = |\nu-\mu| \leq Q$.

The following will be expressed in terms of the type of an irrational number.

Definition. Let g be a positive non-decreasing function at least equal to 1. The irrational α *is said to be of type* $\leq g$ *if for all sufficiently large* B *there exists a solution to*

$$|q\alpha - p| < 1/q \quad \text{and} \quad \frac{B}{g(B)} \leq q < B \tag{2.3.6}$$

where p *and* q *are relatively prime integers.*

The reason why the notion of type will be helpful is that it makes it possible to limit the accuracy with which an irrational number can be approximated by rationals.

Theorem. Expand α *in continued fractions and let* f *be a non-decreasing function at least equal to 1. Assume that the principal convergents* p_n/q_n *satisfy*

$$1/q_n f(q_n) \leq |q_n \ -p_n| \tag{2.3.7}$$

for all sufficiently large n. *Then* α *must be of type* $\leq f$.

Proof: The proof is almost immediate since $|q_n\alpha - p_n| < 1/q_{n+1}$, so that

$$q_{n+1} < q_n f(q_n) \tag{2.3.8}$$

For any large B we can find an n such that

$$q_n < B \leq q_{n+1} \tag{2.3.9}$$

which implies

$$\frac{B}{f(B)} < \frac{q_{n+1}}{f(q_n)} < q_n < B \tag{2.3.10}$$

so that α's type is at most f.

The determination of the type of a given α is seldom easy. One of the many beautiful results in this direction is the following. A number α is of constant type c if and only if the a_i in the continued fraction expansion $\alpha = [a_0, a_1, a_2, \ldots]$ are bounded by c. If α is a quadratic irrational (that is, if it satisfies a quadratic equation with integer coefficients), then it has a periodic expansion in continued fractions so that a_i are uniformly bounded; it is therefore of constant type.

Another sort of result is the metric one where we make statements not about an individual irrational but about almost all.

<u>Theorem</u>. <u>Let</u> h <u>be a positive function with</u>

$$\sum_{n=1}^{\infty} h(n) < \infty \tag{2.3.11}$$

<u>The inequality</u>

$$|q\alpha - p| < h(q) \tag{2.3.12}$$

<u>has then only a finite number of solutions for almost all</u> α.

<u>Proof</u>: Choose a positive ε and find an N such that $\sum_N^\infty h(n) < \varepsilon$; this is possible because of (2.3.11). If α is such that there are infinitely many solutions to (2.3.12), so that $q > N$, then

$$\left|\alpha - \frac{p}{q}\right| < \frac{h(q)}{q} \tag{2.3.13}$$

and α is contained in at least one of the intervals

$$I_\nu = \left(\frac{\nu - h(q)}{q}, \frac{\nu + h(q)}{q}\right) \qquad \nu = 0, 1, \ldots, q-1 \tag{2.3.14}$$

But the union of these intervals has a measure dominated by

$$\sum_N^\infty q \cdot \frac{2h(q)}{q} < 2\varepsilon \tag{2.3.15}$$

As a corollary we can choose $h(n) = 1/[n(\log n)^{1+\delta}]$ with $\delta > 0$. This implies, using (2.3.7), that almost every irrational α is of type $f = (\log n)^{1+\delta}$.

Return to the study of (2.3.1). We shall assume, to begin with, that α is of type $\leq f$ with an unspecified f. Approximate α by a rational p/q (where we of course choose p and q relatively prime to each other). We can then satisfy the inequalities (however large N is)

$$|q\alpha - p| < 1/q$$

(2.3.16)

$$\frac{N}{f(N)} \leq q < N$$

so that

$$\alpha = \frac{p}{q} + \frac{\varepsilon}{q^2}, \qquad |\varepsilon| \leq 1 \tag{2.3.17}$$

Write the partial sums as

$$S_N = S_{N-q} = \sum_{n=N-q+1}^{N} [\{n\alpha\} - \tfrac{1}{2}] = \sum_{m=0}^{q-1} [\{N\alpha - m\tfrac{p}{q} - \tfrac{m\varepsilon}{q^2}\} - \tfrac{1}{2}] \tag{2.3.18}$$

To bound this expression, consider first the influence of the error term $m\varepsilon/q^2$. The function $\{x+b\}$, $a < b < 1$, of x has a discontinuity at $x = 1-b$; otherwise it is continuous with the derivative equal to 1. The influence of the error term is therefore bounded by $q(m\varepsilon/q^2) \leq 1$ plus a finite number of bounded terms corresponding to the jump of the function. The q points $\{\frac{mp}{q}\}$, $m = 0,1,2,\ldots,q-1$, take all the values ν/q, $\nu = Q,1,2,\ldots,q-1$ except possibly for a reordering. Indeed, assume that $\{\frac{mp}{q}\} = \{\frac{m'p}{q}\}$ with $m \neq m'$. Then $(m-m')p = q$ is an integer. But p and q are relatively prime and $|m-m'| < q$, so that $m = m'$. The expression in (2.3.18) is therefore bounded by

$$O(1) + \sum_{\nu=0}^{q-1} [\{\beta - \tfrac{\nu}{q}\} - \tfrac{1}{2}] \tag{2.3.19}$$

An elementary calculation shows that the sum is also $O(1)$. If we assume that $f(x)/x$ is non-increasing, then

$$\int_{N-q}^{N} \frac{f(x)}{x}\,dx \geq q\,\frac{f(N)}{N} \geq 1 \tag{2.3.20}$$

so that we can find an absolute constant K such that

$$S_N - S_{N-q} \leq K \int_{N-q}^{N} \frac{f(x)}{x}\,dx \tag{2.3.21}$$

or, repeating the same argument,

$$S_N = O\left(\int_{1}^{N} \frac{f(x)}{x}\,dx\right) \tag{2.3.22}$$

This proves

<u>Theorem</u>. <u>If the irrational number</u> α <u>is of type</u> $\leq f$ <u>and if</u> $f(x)/x$ <u>is decreasing</u>,

then (2.3.22) holds.

Corollary. Almost all α are of type $\leq (\log x)^{1+\delta}$. Let instead α be a quadratic irrational so that it is of constant type f = c. Then (2.3.22) gives us the bound

$$\frac{1}{N} S_N = O(\frac{\log N}{N}) \tag{2.3.24}$$

Now let us replace the special function $\{x\} - 1/2$ by a function f that is assumed to satisfy some regularity conditions but otherwise is general (see ref. 3).

Theorem. Assume that f is periodic on [0,1] and has three continuous derivatives. If α is a quadratic irrational then

$$I_N^* - I = O(\frac{1}{N}) \tag{2.3.25}$$

with

$$I = \int_0^1 f(x)\, dx$$

$$I_N^* = \frac{1}{N} \sum_{n=1}^{N} f(\{n\alpha\}) \tag{2.3.26}$$

Proof: Because of the regularity assumptions we can write

$$f(x) = \sum_{\nu=-\infty}^{\infty} f_\nu e^{2\pi i \nu x} \tag{2.3.27}$$

with $f_\nu = o(|\nu|^{-3})$. This bound on the Fourier coefficients

$$f_\nu = \int_0^1 f(x)\, e^{-2\pi i \nu x}\, dx \tag{2.3.28}$$

is directly obtained by integrating (2.3.28) by parts three times and applying the Riemann-Lebesgue lemma. We have, since the Fourier series is uniformly convergent, by rearrangement

$$\frac{1}{N} \sum_{n=1}^{N} f(\{n\alpha\}) = \sum_{\nu=-\infty}^{\infty} f_\nu \frac{1}{N} \frac{e^{2\pi i N \nu\alpha} - 1}{e^{2\pi i \nu\alpha} - 1} e^{2\pi i \nu\alpha} \tag{2.3.29}$$

with the obvious interpretation of the term with the $\nu = 0$. Hence

$$|I_N^* - I| \leq \frac{1}{N} \sum_{\nu=-\infty}^{\infty}{}' |f_\nu| \left|\frac{\sin N\nu\alpha}{\sin \pi\nu\alpha}\right| \tag{2.3.30}$$

where the prime denotes that the term with $\nu = 0$ has been left out. To bound the trigonometric expression we note that $\sin x/x \leq 2$ in $0 \leq x \leq 1/2$ and $\sin x/(1-x) \geq 2$ in $1/2 \leq x \leq 1$, so that

(2.3.31) $$\frac{|\sin x|}{||x||} \geq 2$$

if $||x||$ denotes the distance from x to the nearest integer. Therefore (2.3.30) leads to the bound

(2.3.32) $$|I_N^* - I| \leq \frac{1}{2N} \sum_{\nu=-\infty}^{\infty}{}' |f_\nu| \frac{1}{||\nu\alpha||}$$

To bound $||\nu\alpha||$ away from zero we could appeal to what has been said about diophantine approximation. What we need can easily be derived directly, however. Indeed, say that $\alpha = (k+\ell\sqrt{m})/j$, with m not a square, and introduce the conjugate quadratic irrational $\alpha' = (k-\ell\sqrt{m})/j$. Assume that there are infinitely many integer solutions q_1, p_1, q_2, p_2 of the inequality

(2.3.33) $$|\alpha - \frac{p}{q}| < c/q^2$$

But

(2.3.34) $$\begin{aligned}(q_n\alpha - p_n)(q_n\alpha' - p_n) &= (q_nk - jp_n + q_n\ell\sqrt{m})(q_nk - jp_n - q_n\ell\sqrt{m})/j^2 \\ &= \frac{(q_nk - jp_n)^2 - q_n^2\ell^2 m}{j^2} = \frac{\text{non-zero integer}}{j^2}\end{aligned}$$

so that

$$|q_n\alpha' - p_n| \geq \frac{1}{j^2|q_n\alpha - p_n|} > \frac{q_n}{cj^2}$$

which shows that we must have

(2.3.35) $$c > \frac{q_n}{j^2|q_n\alpha' - p_n|} = \frac{1}{j^2|\alpha' - (p_n/q_n)|} > \frac{1}{2j^2|\alpha' - \alpha|}$$

for large values of n.

Choose a value of c that violates (2.3.35); then (2.3.33) is false except possibly for a finite number of q's. Hence there is a $c' > 0$ such that

(2.3.36) $$|q\alpha - p| > c'/q\ , \qquad q = 1,2,\ldots$$

Applying this to (2.3.32), we get

$$|I_N^* - I_N| \leq \frac{1}{2N} \sum_{\nu=-\infty}^{\infty}{}' \; |f_\nu| \cdot |\nu| \tag{2.3.37}$$

using the bound $o(|\nu|^{-3})$ for the Fourier coefficients; this completes the proof.*

The last two theorems indicate that the pseudo-random sequence obtained from the additive congruence rule, although giving a poor approximation to randomness, may nevertheless be acceptable for Monte Carlo. The resulting error is smaller than the one for truly random sequences. In statistical terminology this means: systematic sampling can be better than random sampling. There are several open questions that should be examined:

a) What happens when we integrate in higher-dimensional space?

b) How about the other methods for generating pseudo-random numbers, especially the multiplicative and mixed congruence methods?

c) Can we get "optimal" pseudo-random sequences for integrating "arbitrary" functions? (This will be reformulated in two different ways later on.)

d) In the computer we work with discrete arithmetic, not with real numbers as in the last theorems. How can we translate the results to discrete form?

This opens up a promising area of research in which our knowledge is at present insufficient.

2.4 Improved Monte Carlo.

The straightforward application of Monte Carlo is sometimes not good enough because of its low accuracy, and can be a waste of computing power. Simple Monte Carlo may be adequate, however, if we do a small-scale exploratory study and do not care very much about high precision. It is different when we go to production runs and have to be concerned about computational economy. Then it usually pays to plan the experiment more carefully, using the ideas discussed below or other refinements. There are several ways of increasing the computational accuracy based on a few simple ideas.

a) Try to do as much as possible (or, rather, as much as is economically defensible) by closed-form or numerical integration; supplement it by Monte Carlo to get the desired integral.

b) Distribute your computing effort with some thought; certain parts of the region of integration may be more important than others.

c) Try to balance your design so that the random elements you introduce are counteracted by other random elements.

Let us now apply this in a concrete way to the problem of evaluating I:

$$I(f) = \int_0^1 f(x)\,dx \tag{2.4.1}$$

The straightforward estimate (2.1.2) gives us a mean square error $\sigma(f)/\sqrt{n}$. Suppose that it is possible to get $I(f_a)$ with high accuracy for some function f_a. We could then use the unbiased estimate

$$I^{**}(f) = I(f_a) + I_n^*(f-f_a) \tag{2.4.2}$$

and get the mean square error

$$E[I^{**}(f) - I(f)]^2 = \frac{\sigma^2(f-f_a)}{n} \tag{2.4.3}$$

so that we have achieved a reduction in error if $\sigma(f-f_a) \leq \sigma(f)$.

A typical case is obtained when we use a linear parametric form

$$f_a(x) = \alpha + \beta g(x) \tag{2.4.4}$$

where α and β are constants to be determined and g is a function whose integral we know. Then

$$\min_{\alpha,\beta} \sigma^2(f-f_a) = \sigma^2(f)[I - \rho^2(f,g)] \tag{2.4.5}$$

where $\rho(f,g)$ is the formal correlation coefficient

$$\rho(f,g) = \frac{I(fg) - I(f)I(g)}{\sigma(f)\sigma(g)} \tag{2.4.6}$$

If we can find a g that is well correlated with f, then (2.4.5) indicates that we have achieved a reduction of variance. It might seem tempting to achieve the minimum of (2.4.5) but this is usually computationally unsound. Indeed, the determination of the corresponding values of α and β requires the computation of integrals at least as complicated as I(f). There are at least two possible ways out of this dilemma. One is to use the same sequences $\{x_i;\ i=1,2,\ldots,n\}$ also to estimate the latter integrals. We would then use

$$I(fg) = \int_0^1 f(x)g(x)dx \sim \frac{1}{n}\sum_1^n f(x_i)g(x_i) \tag{2.4.7}$$

and so on for the other integrals needed to solve (2.4.5). The other way out is to be satisfied with a choice α,β that is not optimal but may be good enough to lower the error a good deal. (Pick values that seem intuitively reasonable!)

Assignments. 1. Look at the integral (2.2.1). Take f_a as a piecewise linear function (two or three pieces may be enough) and program the procedure of the last section. Study what gain in accuracy you get for the same sample size n.

We may be able to see *a priori* that an integral like (2.4.1) gets more important contributions from some part of the integration region than from some other part. Even though our *a priori* knowledge is not given in precise form it may pay off to use it in the design of the Monte Carlo experiment.

Stratify the interval (0,1) into the union of the intervals $(a_\nu, a_{\nu+1})$, $\nu = 0,1,2,\ldots,r$ where $a_0 = 0$ and $a_{r+1} = 1$. Introduce the integrals

$$I_\nu^* = \frac{1}{a_{\nu+1} - a_\nu}\int_{a_\nu}^{a_{\nu+1}} f(x)dx, \quad \nu = 0,1,\ldots,r \tag{2.4.8}$$

Now use n_ν random numbers from the distribution $R(a_\nu, a_{\nu+1})$ and get a straightforward Monte Carlo estimate I_ν^* of I_ν. The estimate

$$I_{strat}^* = \sum_{\nu=0}^{r} (a_{\nu+1} - a_\nu)\, I_\nu^* \tag{2.4.9}$$

is then unbiased with variance

$$\mathrm{Var}(I_{strat}^*) = \sum_{\nu=0}^{r} \frac{(a_{\nu+1}-a_\nu)^2}{n}\, \sigma_\nu^2(f) \tag{2.4.10}$$

where σ_ν^2 stands for the conditional "variance" of f over $(a_\nu, a_{\nu+1})$. If the total number $n = n_0+n_1+\ldots+n_r$ is fixed we ought to distribute the n_ν so that we have approximately

$$n_\nu = \frac{(a_{\nu+1}-a_\nu)}{\sum_{\nu=0}^{r} \sigma_\nu (a_{\nu+1}-a_\nu)}\, n \tag{2.4.11}$$

where we have treated the integers n_ν as continuous quantities. It is usually impractical to use (2.4.11) directly, since the σ_ν's are not available. Indeed, (2.4.11) will serve only as an indication of the manner in which we can guess good (but not optimal) values for n_ν.

This supposes that we have fixed the subdivision points a_ν. How to make this choice is a good deal more difficult, and will not be dealt with here.

2. Return to the integral (2.2.1). Stratify the region of integration into a few intervals. Apply the above method to the evaluation of the integral. Compare the result and efficiency of computation with what you have done in earlier assignments. One should also try to exploit qualitative properties of the integrand. To be concrete, assume that the integrand f in (2.1.1) is monotonically increasing or monotonically decreasing. Then it seems promising to use the fact that of the two functions f(x) and f(1-x) one increases and the other decreases so that they are negatively correlated (prove that they are). Therefore, put

(2.4.12)
$$I^* = \frac{1}{2N}\sum_1^n [f(x_i) + f(1-x_i)]$$

which is an unbiased estimate of I(f) with the variance

(2.4.13)
$$\mathrm{Var}(I^*) = \frac{1}{4n}[2\sigma^2(f) + 2\rho\sigma^2(f)]$$

where ρ is the formal coefficient of correlation between f(x) and f(x-1). Since $\rho < 0$ the expression in (2.4.13) is less in value than that in (2.1.3).

This is the simplest version of another fairly general method of increasing the computational efficiency of Monte Carlo experiments: antithetic variables.

To understand the idea behind the technique of antithetic variables better (see ref. 4), let us look at the stochastic variable

(2.4.14)
$$F^* = \frac{1}{2}[f(x_1) + f(x_2)]$$

To simplify the discussion it will be carried on for the discrete version of the integral. Assuming that both x_1 and x_2 have a uniform discrete distribution over the values $x = 1/n, 2/n, \ldots, n/n$ (condition c), the expression in (2.4.14) is an unbiased estimate of the quantity $F = \frac{1}{n}\sum_1^n f(\frac{\nu}{n})$. Now let us make its variance as small as possible without violating condition c, which only makes a statement about marginal distributions but not about the form of dependence between x_1 and x_2. It is not difficult to see from (2.4.13) that this leads us to the extremum problem

$$\inf_{c} \sum_{i,j=1}^{n} f(\tfrac{i}{n})\, f(\tfrac{j}{n})\, p_{ij} \tag{2.4.15}$$

where the p_{ij} are varied subject to the condition c (the reader will recognize this as the discrete analogue to the integration problem discussed earlier). Condition c means that

$$\sum_{i=1}^{n} p_{ij} = \sum_{j=1}^{n} p_{ij} = 1/n \; ; \quad \text{all } i,j. \tag{2.4.16}$$

This implies that if we consider the numbers p_{ij} as coordinates in n^2-space we will be in a region where these coordinates satisfy (2.4.16) and $p_{ij} \geq 0$. This means that c is a convex, closed polyhedron whose vertices correspond to $p_{ij} = 0$ or $1/n$; the p_{ij} then form, because of (2.4.16), a permutation matrix up to a multiplicative constant. For a given $P = \{p_{ij}\}$ we can write

$$P = \sum_{k=1}^{n!} w_k \, P^{(k)} \tag{2.4.17}$$

where the $P^{(k)}$ are the n! permutation matrices and the weights w_k are non-negative. This is geometrically equivalent to representing the convex polyhedron as the convex hull of its vertices. Because of (2.4.17) the problem (2.4.15) reduces to

$$\min_{P^{(k)}} \sum f(\tfrac{i}{n})\, f(\tfrac{j_i}{n}) \tag{2.4.18}$$

where j_i is the permutation of $1,2,\ldots,n$ corresponding to the permutation matrix $P^{(k)}$. This is a familiar problem: rearrangement of series. We look for that rearrangement making the two sequences $\{f(\frac{i}{n})\}$ and $\{f(\frac{j_i}{n})\}$ monotonic in opposite order (see ref. 5, chapter 10).

Note that the stochastic dependence of the probability distribution described by $\{P_{ij}\}$ actually degenerates into functional dependence in the solution given above. Now this was for a very special case, but qualitatively the same holds in more general situations (see ref. 4).

It is this sort of functional dependence which we use in the technique of antithetic variables in order to reduce the variance of the estimate.

Assignments. 1. Using antithetic variables, evaluate the integral in (2.2.1) by Monte Carlo. Compare the computational efficiency with that of other methods. Discuss combinations of the basic ideas we have considered.

2. Compare the computational efficiency of the various ways of estimating the integral (2.2.1) by counting the number of elementary arithmetic operations needed for each method. Compare this with the resulting variance to judge their relative merits.

When using simulation in an operations research problem we are usually confronted by a system, say $S(\alpha)$, whose complexity is so considerable that a purely analytic treatment is impossible or uneconomical. Here α represents a controlled parameter, a real number or vector, upon whose value we decide. The parameter may, for example, describe some variable in a production process, or a capacity in a network. We would like to know how a given criterion $C(\alpha)$, expressing the overall performance of $S(\alpha)$, varies with α. To find this out we pick some values for the parameter, say $\alpha_1,\alpha_2,\ldots,\alpha_p$, and make a Monte Carlo experiment for each α_ν. We get, in this way, p estimates $C_1^*,C_2^*,\ldots,C_p^*$ of the quantities $C(\alpha_1),C(\alpha_2),\ldots,C(\alpha_p)$. Let us assume that the estimates are unbiased with variances $\sigma_1^2,\sigma_2^2,\ldots,\sigma_p^2$.

After having done this, we plot the values C_i^* against α to study the relation between the parameter and the output. Note that we can write

$$C^*(\alpha) \simeq C(\alpha) + N(0,\sigma_\alpha) \tag{2.4.19}$$

if the sample size of the simulation experiment is large enough; σ_α is inversely proportional to $\sqrt{n_\alpha}$ where n_α is the sample size used for the value α. In particular, if $C(\alpha)$ has some simple form, such as the linear one $C(\alpha) = a + b\alpha$, we can use the C_i^* first to estimate a and b and then $C(\alpha)$ for other values than the α, or for a particular one of them. In the latter case we get a reduced variance that can be considerably smaller than σ_α. We just use ordinary linear regression analysis (see, for instance, the last chapter of ref. 7). Similarly, if $C(\alpha)$ is a quadratic polynomial in α we can estimate for instance the location of a maximum or minimum of $C(\alpha)$.

Assignment. Consider the integral

$$C(\alpha) = \alpha \int_0^1 e^{-\alpha/x}\,dx, \quad \alpha > 0$$

and design a Monte Carlo experiment to find the approximate location of the value α_0 maximizing $C(\alpha)$.

2.5 Quadrature.

For simplicity the assignments in this section have been given on the real line but the real advantages of the Monte Carlo method become apparent only when the dimensionality is high, say 10 or more. In later chapters we shall apply the method extensively to such cases.

Recalling the theoretical discussion in section 2.3, it is at this point natural to ask how we should judge the performance of the method when pseudo-random numbers are used, for instance when one of the congruence generators discussed in chapter 1 is applied.

The natural attitude is to look at Monte Carlo as a method of numerical quadrature. Assume that the integration is performed in the r-dimensional unit cube C so that $X = (X_1,X_2,\ldots,X_r)$ with $0 \leq X_\nu \leq 1$; $\nu = 1,2,\ldots,r$. We would write a quadrature formula quite generally as

$$I^*(f) = \sum_{\mu=1}^{n} C_\mu f(X^\mu) \tag{2.5.1}$$

so that the problem consists in choosing the points $X^\mu \subset R^r$ and the coefficients C_μ in such a manner that the error of approximation $I-I^*$ becomes small. The set of X^μ determines the type of quadrature design. The simplest design consists of choosing a set of m points $\xi_1,\xi_2,\ldots,\xi_m$ in (0,1) and using all the points of the form $(\xi_{i_1}, \xi_{i_2},\ldots,\xi_{i_r})$ for the X^μ's. This is a factorial design with the number of X^μ's equal to m^r. In such a design the problem has been reduced to the choice of a one-dimensional array of numbers together with their coefficients. Already for moderate values of n the number m^r becomes prohibitively large, and we must turn to non-factorial designs.

It seems reasonable to ask that the X^μ's be uniformly spaced throughout C (which could be given a meaning for a factorial integration design), but to give this a precise meaning presents some difficulty when we cannot make asymptotic statements; the latter are meaningless when the number n of points X_ν is so small that $\frac{1}{r} \ln n$ is, say, less than one.

An ambitious attack on this problem would start as follows. Assume that we are looking for an integration design and a quadrature expression for some class F of integrands f. The class F is usually given to us in a qualitative way by presenting smoothness through a continuity modulus, perhaps a Lipschitz condition or by prescribing the existence and continuity of a number of derivatives. We may also have direct restrictions on the size of the integrand at various points of the domain of integration, occasionally supplemented by conditions like positivity or convexity.

One way of expressing such restrictions is to give a probability measure over our function space indicating the likelihood (possibly interpreted in a Bayesian manner) of various integrands. It goes without saying that such a probability measure m should not be taken too literally: it only serves as a qualitative indication of what smoothness properties to expect of the integrand.

The quadrature problem can then be formulated as follows: *find* X^μ *and* C_μ *such that*

$$\min_{X^\mu} \min_{C_\mu} E[I(f) - I^*(f)]^2 \tag{2.5.2}$$

is realized, at least approximately.

To make the discussion more concrete consider the following special case of (2.5.2); actually, it has some general consequences which will help us in the further study of the problem. It is also of direct interest for the computational treatment of stochastic integrals and stochastic differential equations dealt with in later chapters.

Choose m as the probability measure in function space corresponding to the Wiener process so that

$$\begin{aligned} &E\, f(x) \equiv 0 \\ &E\, f(x)\, f(y) = \min(x,y) \end{aligned} \tag{2.5.3}$$

We can then write

$$I(f) = \int_0^1 f(x)dx = \int_0^1 (1-x)f(dx) \quad ; \quad I^*(f) = \sum_1^n c_\nu f(x_\nu) = \int_0^1 \phi(x)\, f(dx) \tag{2.5.4}$$

where ϕ is a step function, constant except for the abscissa x_ν where it has a jump $-c_\nu$.

Before dealing with (2.5.2) we shall compare the performance of some of the standard quadrature formulae, starting with Simpson's rule

(2.5.5) $$I_s^*(f) = \frac{1}{3n} f(0) + \frac{4}{3n} f(\frac{1}{n}) + \frac{2}{3n} f(\frac{2}{n}) + \frac{4}{3n} f(\frac{3}{n}) + \ldots$$

(the first term vanishes identically in our case). Hence ϕ is given by

(2.5.6) $$\phi(x) = \begin{cases} \phi_1 = 1 - \frac{1}{3n} & 0 < x < \frac{1}{n} \\ \phi_2 = 1 - \frac{5}{3n} & \frac{1}{n} < x < \frac{2}{n} \\ \phi_3 = 1 - \frac{7}{3n} & \frac{2}{n} < x < \frac{3}{n} \\ \phi_4 = 1 - \frac{11}{3n} & \frac{3}{n} < x < \frac{4}{n} \\ \phi_5 = 1 - \frac{13}{3n} & \frac{4}{n} < x < \frac{5}{n} \\ \text{etc.} \end{cases}$$

Using (2.5.5) and (2.5.6) gives us

(2.5.7) $$E[I(f) - I_s^*(f)]^2 = \int_0^1 [1-x-\phi(x)]^2 dx = \sum_{\nu=1}^{n} \int_{(\nu-1)/n}^{\nu/n} [1-x-\phi_\nu]^2 dx$$

$$= \frac{1}{3} \sum_{\nu=1}^{n} [(1 - \frac{\nu}{n} + \phi_\nu)^3 - (1 - \frac{\nu}{n} - \phi_\nu)^3]$$

Because of (2.5.4), this reduces to

(2.5.8) $$E[I(f) - I_s^*(f)]^2 = \frac{1}{3}[(\frac{1}{3n})^3 + (\frac{2}{3n})^3 + (\frac{2}{3n})^3 + (\frac{1}{3n})^3 + (\frac{1}{3n})^3 + (\frac{2}{3n})^3 + \ldots]$$

so that Simpson's rule performs as described by its mean square error

(2.5.9) $$E[I(f) - I_s^*(f)]^2 \sim \frac{1}{9n^2}$$

(How good is this?) To find the minimum (but so far keeping the $X^\mu = \frac{\mu}{n}$), we see from (2.5.7) (leaving out details) that we should put

(2.5.10) $$\phi_\nu = 1 - \frac{\nu}{n} + \frac{1}{2n}$$

This gives us the quadratic error

(2.5.11) $$E[I(f) - I^*(f)]^2 \sim \frac{1}{12n^2}$$

We thus have the slightly paradoxical result that Simpson's rule is 33% worse than the crude quadrature formula with the weights (2.5.10). It is also possible to verify that, in the present case, equidistant X^μ's are best.

This case is more general than it may appear. First, the assumption that f(x) form a Wiener process was used only as far as second-order properties are concerned, not the fact that this assumption implies normality. It is therefore enough to assume that f(x) is an orthogonal process with homogeneous differential variance. Secondly, if f(x) has some other probabilistic structure it is often possible to reduce its integration to a problem similar to the one we have just dealt with. Indeed, let us recall the manner in which we approach the question of model building for random phenomena.

We are governed by two principles in order to simplify or even make possible specification of the whole simultaneous probability distribution. The first is to reduce the complexity of the phenomenon by breaking it up into pieces all of which can be treated as independent of each other. In this way we have only to specify the marginal distributions. The second guiding principle is to look for invariances, in time or space or with respect to transformations of the phenomenon, which leave the probabilities of the system unchanged. We try in this way to limit the form of the marginal distributions as much as possible. Successful application of these two principles requires good intuitive understanding of the qualitative functioning of the system.

A typical example of this procedure is the study of the distribution properties of some noise source. We try to approximate the source by regarding it as the output of an RC-filter with white noise as input, which means that the output is an exponential average over the input. Behind this assumption lies the hypothesis that the input exhibits complete independence between the events over disjoint time intervals, and the hypothesis that the distributions of the white noise are invariant over time. We thus get the noise represented as

$$n(t) = \alpha \int_{-\infty}^{t} e^{-\beta(t-s)}\, w(ds) \tag{2.5.12}$$

and are left with the determination of three parameters, α, β, and the variance per unit time of w(t). These parameters have to be estimated from data. We get similar although more complicated expressions than (2.5.12) if we assume other passive filters, but the important assumption is the linear structure. The quadrature problem for n(t) can then be reformulated as the numerical determination of the integral

(integrating by parts)

$$j = \int g(x)\, W(dx) = \int G(x)\, W(x)\, dx \tag{2.5.13}$$

over some domain. Our approximations can now be written as

$$j^* = \sum c^{\mu}\, G(x^{\mu})\, W(x^{\mu}) = \int \gamma(x)\, W(dx) \tag{2.5.14}$$

where γ is a step function with the steps at x^{μ} and the previous integrand $f(x)$ is now replaced by $G(x)\, W(x)$. The mean square error is then, in the same manner as before, found as

$$E[j-j^*] = \text{const.} \int [G(x) - \gamma(x)]^2\, dx \tag{2.5.15}$$

(Compare with (2.5.7)!) The best choice of coefficients corresponds to giving γ the value

$$\frac{1}{x^{\mu+1}-x^{\mu}} \int_{x^{\mu}}^{x^{\mu+1}} G(x)\, dx \tag{2.5.16}$$

in the interval $(x^{\mu}, x^{\mu+1})$, and (2.5.15) then reduces to

$$E[j-j^*] = \text{const.} \sum (x^{\mu+1}-x^{\mu})\, \gamma_{\mu}^2 \tag{2.5.17}$$

where γ_{μ}^2 is the "conditional variance" over $(x^{\mu}, x^{\mu+1})$. This is no longer uniform so that the best spacing is not uniform. One should, in a chapter on sample surveys, return to the determination of the x^{μ}, the integration design, because of the formal similarity of the present problem to that of optimum stratification. It is enough just now to realize that the choice of weights is fairly simple (see ref. 18).

To summarize, we have found that for this type of fairly chaotic integrand we should stay away from the more sophisticated quadrature methods, since we actually lose by using them and, moreover, they often require more computing time than the cruder methods.

This study should be extended to high-dimensional cases since it is these in which we are interested for Monte Carlo (but not necessarily for stochastic differential equations). This has not yet been done but presents a promising area of research.

A discussion of this problem is illuminated by supplementing the above approach by another, less Bayesian one, where F is given by conditions of the Lipschitz type:

$$(2.5.18) \qquad |f(x) - f(x')| \leq K||x - x'||, \quad x \varepsilon R^r$$

We can reformulate the problem by replacing (2.5.2) by

$$(2.5.19) \qquad \min_{C_\nu, x^\mu} \max_{f \varepsilon F} |I(f) - \sum C_\mu f(x^\mu)|$$

(Discuss (2.5.19) at least in the one-dimensional case and compare with the earlier discussion!)

2.6 Conclusions.

We have seen how to apply the Monte Carlo method not only in its simplest form but also in refined versions. It was observed that we can achieve considerable savings in computation time by applying some rather simple ideas for variance reduction. This will be useful to us in some of the later chapters of this book.

We have also seen that the use of pseudo-random numbers in stochastic simulation may be more advantageous than may appear likely at first glance. A few special techniques of generating pseudo-random sequences tailored to particular distributions were examined.

Our discussion led us naturally to consider the performance of Monte Carlo from the point of view of numerical quadrature. Our knowledge of this subject is still very incomplete but we found as a side result how to deal with computational problems arising in the evaluation of stochastic integrals and in the solution of stochastic integral equations, as we shall be led to do repeatedly later on.

CHAPTER 3:

LIMIT THEOREMS

3.1 Limits of Convolutions.

Among the most useful results of probability theory are the probabilistic limit theorems, and we shall devote this chapter to the examination of a few of them from the computational viewpoint.

Consider a sample $x_1, x_2, \ldots, x_n$ drawn from a probability distribution with cumulative distribution function $F(x)$. Form the sum

$$S = S_n = x_1 + x_2 + x_3 + \ldots + x_n \tag{3.1.1}$$

and study its behavior when n takes large values. Assume that the x's are stochastically independent.

To get some feeling for the concrete nature of the problem, we shall start from some special cases that are intended to illustrate the general situation.

Assignments. 1. Let F be the distribution function corresponding to a Bernoulli variable

$$F(x) = \begin{cases} 0 & \text{if } x < 0 \\ 1-p & \text{if } 0 < x < 1 \\ 1 & \text{if } 1 < x \end{cases} \tag{3.1.2}$$

so that

$$P(x_i=1) = p \quad \text{and} \quad P(x_i=0) = 1 - p = q \tag{3.1.3}$$

Simulate the sample on the machine for n = 5, 10, 15, ..., 50; calculate the empirical distribution function

$$F_N^*(x) = (\text{number of times that } T \leq x)/N \tag{3.1.4}$$

where N is the number of replications in each simulation and T is the standardized sum

$$T = \frac{S - np}{\sqrt{npq}} \tag{3.1.5}$$

Plot the empirical distribution functions, for instance by using the APL plotting routine. Choose values of p in the interval (0.2,0.8).

2. Do the same assignment using values of p from the interval (0,0.2) or (0.8, 1.0). Compute $F_N^*(x)$ not from T but from the non-standardized sum S (for the first interval) or n-S (for the second interval). Plot the results.

As we know already, both assignments will result in convergence of $F_N^*(x)$, in the first case to the normal distribution function

$$F_N^*(x) \to \frac{1}{\sqrt{2\pi}} \int_{-\infty}^{x} e^{-t^2/2}\, dt \tag{3.1.6}$$

and in the second case to the Poisson distribution function

$$F_N^*(x) \to \sum_{0}^{[x]} e^{-\lambda} \frac{\lambda^k}{k!} ; \quad \lambda = np \tag{3.1.7}$$

Seeing the output from the assignment will help us to get a more concrete feeling for the situation and for the speed of convergence.

If n is not large we can still compute the (theoretical) distribution function for S_n (analytically, for any F) in the following way.

Assume, first, that x_1 has a discrete probability distribution with probabilities $p_1, p_2, \ldots, p_\alpha$ and x_2 has a (possibly different) discrete distribution with probabilities $q_1, q_2, \ldots, q_\beta$; we assume in both cases that the values taken by x_1 and x_2 are the α and β first natural numbers, so that

$$\begin{aligned} &P(x_1 = k) = p_k; && k = 1,2,\ldots,\alpha \\ &P(x_2 = k) = q_k; && k = 1,2,\ldots,\beta \\ &\sum_1^\alpha p_k = \sum_1^\beta q_k = 1 \end{aligned} \tag{3.1.8}$$

We thus get

$$P(x_1 + x_2 = \ell) = \sum_{k=-\infty}^{\infty} P(x_1 = k) \cdot P(x_2 = \ell - k) = \sum p_k q_{\ell-k} \tag{3.1.9}$$

summed over the appropriate range, where both factors of the terms are positive. It is clear that the possible values of $x_1 + x_2$ will range from 2 to $\beta+\alpha$ so that they are $\alpha+\beta-1$ in number. The program could be organized in terms of inner products of the two vectors Q1 and P1, both in $(\alpha+\beta)$-space,

$$Q1 = (q_\beta, q_{\beta-1}, q_{\beta-2}, \ldots, q_2, q_1, \underbrace{0, 0, \ldots 0, 0}_{\alpha})$$

(3.1.10)

$$P1 = (\underbrace{0, 0, 0, \ldots, 0, 0}_{\beta}, p_1, p_2, \ldots p_\alpha)$$

or, in APL notation,

(3.1.11)

```
Q1 ← (⌽Q), (⍴P)⍴0
P1 ← ((⍴Q)⍴0), P
```

Now let R be some vector with $\alpha+\beta-1$ components, e.g.

(3.1.12)

```
R ← (¯1 + (⍴P) + ⍴Q) ⍴0
```

and put

(3.1.13)

```
R[I] ← +/((-I)⌽Q1) × P1
```

for $I = 1,2,\ldots,\alpha+\beta-1$; then the entries in the R-array are the desired probabilities in (3.1.9), or symbolically $R = P * Q$, the convolution between P and Q. (Note, however, that the values of x_1+x_2 have been shifted one step compared to the index of the R-vector.)

In this case it seems wise not to exploit the array-handling capabilities of APL to their full extent, since space limitations may arise if we want to convolve probability distributions with many points.

If we have several terms $x_1, x_2, x_3, \ldots$ and want to find the distribution of their sum, we just call in our program by a loop. In particular, we can find convolution powers $P*P*\ldots*P = P^{n*}$ in this way.

If the x's have continuous probability distributions we first have to discretize and replace the integrals corresponding to (3.1.9) by sums.

Assignment. Let us go back to the random number routine GAUSS discussed in chapter 1. We take $x_1, x_2, \ldots, x_{12}$ to be uniformly distributed over the interval (0,1). Compute, using an APL program for convolutions, the distribution function of

$$Y = x_1 + x_2 + \ldots + x_{12} - 6$$

Compare it with a table of the normal distribution function. What conclusions can we draw concerning the accuracy of the subroutine GAUSS?

To get some feeling for the role played by the assumption that the x's are independent, we shall relax the condition and see what happens.

Let $x_1, x_2, x_3, \ldots$ be generated by a Markov chain with the $r \times r$ probability transition matrix

$$M = \begin{bmatrix} \frac{p}{r}+1-p, & \frac{p}{r}, & \frac{p}{r}, & \cdots & \frac{p}{r} \\ \frac{p}{r}, & \frac{p}{r}+1-p, & \frac{p}{r}, & \cdots & \frac{p}{r} \\ \cdot & \cdot & \cdot & \cdot & \cdot \\ \frac{p}{r}, & \frac{p}{r}, & \frac{p}{r}, & \cdots & \frac{p}{r}+1-p \end{bmatrix} = \{\frac{p}{r} + (1-p)\delta_{ij};\ i,j=1,2,3,\ldots,r\}$$

where r is some small integer, say 3 or 4.

Assignment. Write a simulation program which generates the x-sequence. Pick x_1 each time uniformly distributed over the discrete interval $1,2,3,\ldots,r$, and form the sum $x_1+x_2+\ldots+x_n$ for n = 5, 10, 15, 20. Iterate the procedure, as described in chapter 2, and compare the convergence of the empirical distribution functions $F_n^*(x)$ for the values p = 1, .8, .6, .4, .2, 0. What conclusions do you draw about the influence of the value of p on the behavior of F_n^*? What does the parameter p signify in terms of dependence between the x's?

3.2 An Insurance Model.

The following model has been used to describe one aspect of the business of an insurance company. Let time t take values 1,2,3,..., where we could think of the time unit as a fraction of one day, and associate with this time unit a probability p for the occurrence of exactly one claim for a particular policy out of the portfolio of M policies (we disregard the possibility of more than one claim per time unit). If a claim occurs we denote the distribution function of its size (in some monetary unit) by F.

At the same time the company receives an amount, say P, in premiums paid over a period of one year which should be added to the initial reserve fund U available at the beginning of the year.

Now we will try to get some idea of what the probability of ruin (at the end of the year) is; that is, the value of

$$P\{U + P < \text{total sum of claims}\} = P\{U + P < T\} \qquad (3.2.1)$$

where T stands for the total sum paid out.

To get an approximation to the probability we can reason as follows. Write

$$T = C_1 + C_2 + \dots + C_r \tag{3.2.2}$$

where r is the number of claims and C_1, C_2, etc., are the individual claims. The probability distribution of r is binomial so that, with $N = 365 \cdot M$,

$$P(r=k) = \binom{N}{k} p^k (1-p)^{N-k}, \quad k=0,1,2,\dots,N \tag{3.2.3}$$

(note that this assumes that claims occur independently of each other). If $N_p \gg 1$ and $N_r \ll N$, we know that this probability distribution can be approximated by the normal one. On the other hand, if r is fixed and large, then T also has an approximately normal distribution, due to the central limit theorem.

To make this precise we note that

$$ET = \sum_{r=0}^{N} P(r=k)\, E(T|r=k) = \sum_{r=0}^{N} P(r=k) \cdot r \cdot m = rNp \tag{3.2.4}$$

where we have introduced the average claim

$$m = EC_i = \int x\, F(dx) \tag{3.2.5}$$

In a similar way

$$ET^2 = \sum_{r=0}^{N} P(r=k)\, E(T^2|r=k) = \sum_{r=0}^{N} P(r=k)\, [r\sigma^2 + r^2 m^2] \tag{3.2.6}$$

$$= Np\sigma^2 + (Np(1-p) + N^2 p^2) m^2$$

with

$$\sigma^2 = \mathrm{Var}(C_i) = \int_0^\infty (x-m)^2\, F(dx) \tag{3.2.7}$$

We get

$$\mathrm{Var}(T) = N[p\sigma^2 + m^2 p(1-p)] \tag{3.2.8}$$

where the variability has been split into two parts: the first due to the variation of the size of the claims and the other to the fluctuation of the number of claims. The desired approximation would then be

(3.2.9) $$P(\text{ruin}) = 1 - P\{T \leq U + P\} \simeq 1 - \Phi\left(\frac{U + P - Npr}{(N[p\sigma^2 + m^2 p(1-p)])^{1/2}}\right)$$

(See ref. 19.)

Before discussing the merits of the approximation, let us simulate on the computer this model (due to Filip Lundberg) of the risk process of an insurance activity. It is enough for the present purpose to do this on a small scale, but we should try to represent the result as intuitively as possible. The program LUNDBERG (see Appendix 3, figure 3.1) corresponds to the model above with

FUND = U

LAM = p

T = time parameter

PREMIUM = P

Three realizations are given in Appendix 3 (see figures 3.2 - 3.4). In the last one, ruin is seen to have occurred; in the others, ruin did not occur during the time interval displayed in the plots.

Returning to (3.2.9), let us ask how reliable the approximation is. One way of finding out would be to use the program LUNDBERG (but without the graphical display) for a Monte Carlo experiment and to count the frequency of ruin out of a certain number of trials.

Alternatively, we can approach the problem from the analytical and numerical side. Since

(3.2.10) $$P(T \leq x) = \sum_{k=0}^{N} P(r=k) \cdot P\{C_1+C_2+\ldots+C_k \leq x\}$$

and since sums of independent stochastic variables give rise to convolution powers (note the independence assumption), we get

(3.2.11) $$P(T \leq x) = \sum_{k=0}^{N} \binom{N}{k} p^k(1-p)^{N-k} F^{k*}(x)$$

We can now use one of our programs to calculate the convolution powers for values of k such that the probability of occurrence is appreciable.

Assignment. Carry this out on the computer and compare it with the approximation (3.2.9). Discuss the result and, possibly, alternative approaches. Use small values of N to keep the computing work moderate.

3.3 Approximations.

We have looked into some numerical aspects of convergence to the normal and Poisson distributions. To deepen our understanding of how this convergence takes place we shall look at the behavior of normed partial sums S_t of the form

$$(3.3.1) \qquad M_t = (x_1+x_2+x_3+ \dots + x_t)/t = S_t/t$$

where the terms x_i are independent stochastic variables with some frequency function f. For reasons that will be clear later on it will be illuminating to choose the frequency function as

$$(3.3.2) \qquad f(x) = \begin{cases} \frac{1}{2A} \frac{1}{x^{1+1/A}} ; & |x| \leq 1 \\ 0 ; & |x| > 1 \end{cases}$$

where A is a positive parameter.

We generate a sample from f, form M_t and display it graphically. If u stands for a uniformly distributed stochastic variable, then $v = u^{-a}$ has a distribution function G

$$(3.3.3) \qquad G(x) = P\{v \leq x\} = P\{u^A \geq 1/x\} = P\{u \geq \frac{1}{x^{1/A}}\} = 1 - 1/x^{1/A}$$

if $x \geq 1$, so that the frequency function is

$$(3.3.4) \qquad g(x) = G'(x) = \frac{1}{A} \frac{1}{x^{1+1/A}}$$

Therefore $w = \pm u^{-a}$, with its sign randomized, has the desired frequency function. A possible program is the program TAIL (figure 3.5 of Appendix 3). Since the behavior of the tail is important, one has to discretize the stochastic variable u by small steps, say .0001.

Now it is easy to generate and display the stochastic process M(t), t = 1,2,..., n; see the program CONTPL (figure 3.6 of Appendix 3).

Execute CONTPL for A=1/5 and N=30. The result is shown in figure 3.7 of Appendix 3. For small values of t the function is a bit variable but it tends fairly quickly toward the value m = 0. For A = 1/3 (see figure 3.8) the behavior is similar although less pronounced.

When we go to larger values of A the behavior changes. For A = 1/2 (figure

3.9) the departure is not as drastic, however, as for A = 1.0 (figure 3.10) and it is even more obvious for A = 2 (figure 3.11).

It is easy to explain this change of behavior. Indeed, let us calculate the moments

$$\text{(3.3.7)} \qquad EX^p = \frac{1}{a}\int_1^\infty x^{p-1-1/a}\,dx = \begin{cases} 1/(1-ap) & \text{if } p < 1/a \\ \infty & \text{if } p \geq 1/a \end{cases}$$

Hence for A = 1/5, moments exist up to (integral) order 4; for A = 1/3 moments up to 2. For the value A = 1/2 the mean exists, which is not the case for the remaining cases. The lack of convergence in these later cases simply corresponds to the law of large numbers not being applicable.

The practical consequences of this simple observation are not always realized. An argument often heard goes as follows. In practical situations the variables encountered are always bounded so that moments of all orders exist and we can therefore put our trust in the law of large numbers.

The argument is only superficially correct and is based on a misunderstanding of the question how to interpret mathematical conditions (like "moments should exist") in practical terms. Actually, while there may be an upper bound for the values of our variables, the decrease of the frequency function may be so slow that the mathematical notion of divergence is a better approximation to reality than convergence to a large value. This is the reason for the behavior we have observed; the practical consequences of this should be borne in mind when analyzing real data: we should not believe in the law of large numbers automatically, but should examine critically the validity of conditions necessary to guarantee its applicability. Similar statements hold for other probabilistic limit theorems.

There is another, slightly more sophisticated, conclusion we can draw from our simulation experiment. In figures 3.7 and 3.8 of Appendix 3 the curve M looks continuous, but this is not so for figures 3.9 to 3.11.

In the first two cases the second-order moments exist and we can apply the central limit theorem. Therefore the stochastic function M(t) can be approximated with the Wiener process (to be discussed in chapter 4), which is known to have continuous sample functions. If the variance does not exist the approximation is in terms of some other stochastic process with independent increments. Thus one can have discontinuities; actually the probability of at least one discontinuity tends to 1 if the length of the time interval tends to infinity. This explains the discontinuous behavior in the second three cases.

```
      ∇LUNDBERG[⎕]∇
    ∇ LUNDBERG
[1]    T←1
[2]    (10ρ' '),'0',(24ρ' '),'25',(24ρ' '),'50'
[3]    V←(10ρ'-'),'|',50ρ'-'
[4]    V[11+FUND]←'o'
[5]    V
[6]   LR1:FUND←FUND+PREMIUM
[7]    →(LAM≥0.0001×?10000)/LR2
[8]    W←(10ρ' '),'|',50ρ' '
[9]    W[11+⌊FUND]←'o'
[10]   W
[11]   T←T+1
[12]   →LR1
[13]  LR2:SUM←-M×⍟0.001×?1000
[14]   FUND←FUND-SUM
[15]   →(FUND≥0)/LR3
[16]   'RUIN OCCURRED AT TIME';T
[17]   →0
[18]  LR3:W←(10ρ' '),'|',50ρ' '
[19]   W[11+⌊FUND]←'o'
[20]   W
[21]   T←T+1
[22]   →LR1
    ∇
```

Figure 3.1

As initial values one could choose

FUND ← 25

PREMIUM ← 1

LAM ← .4

M ← 5

and obtain as output figure 3.2.

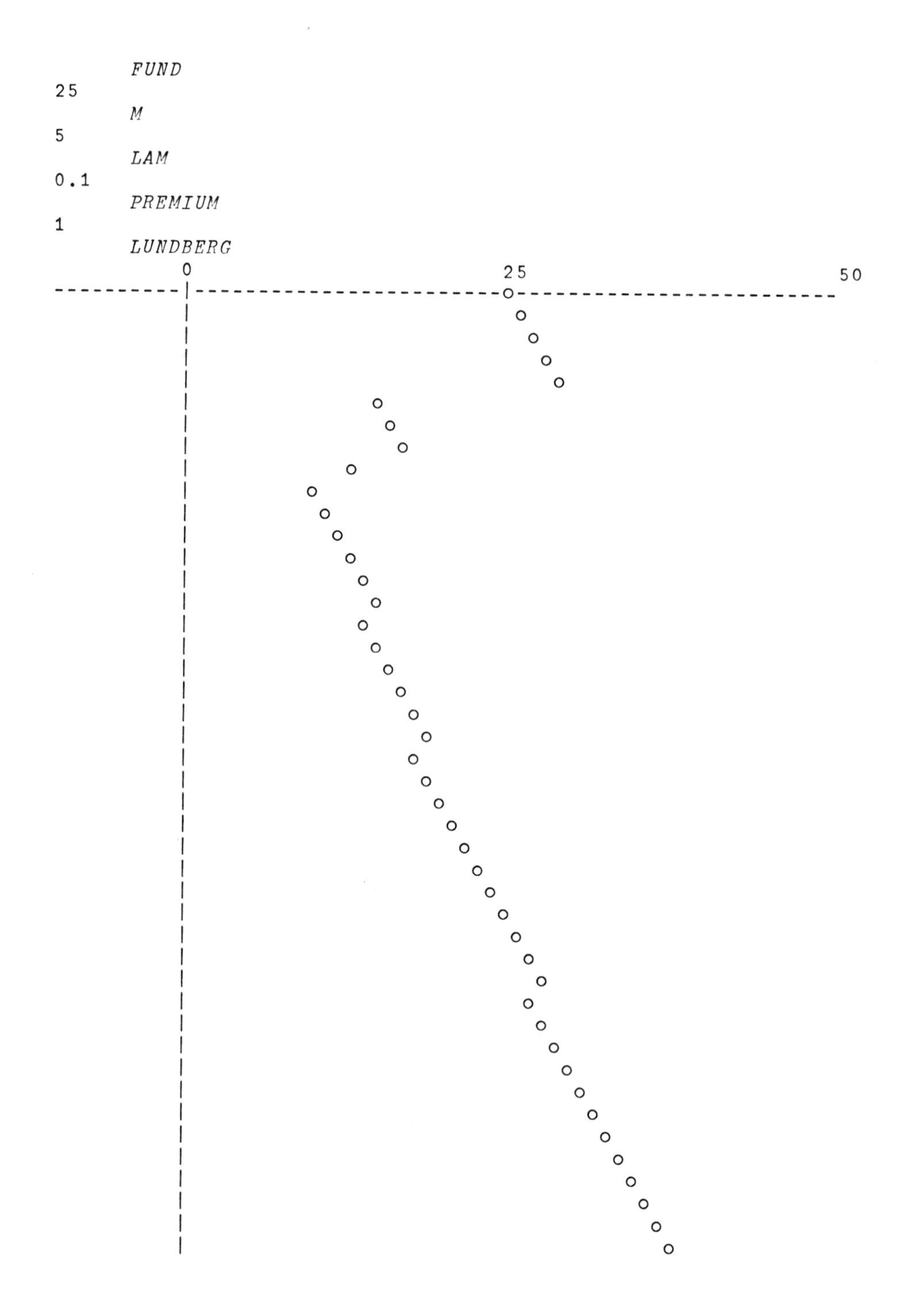

Figure 3.2

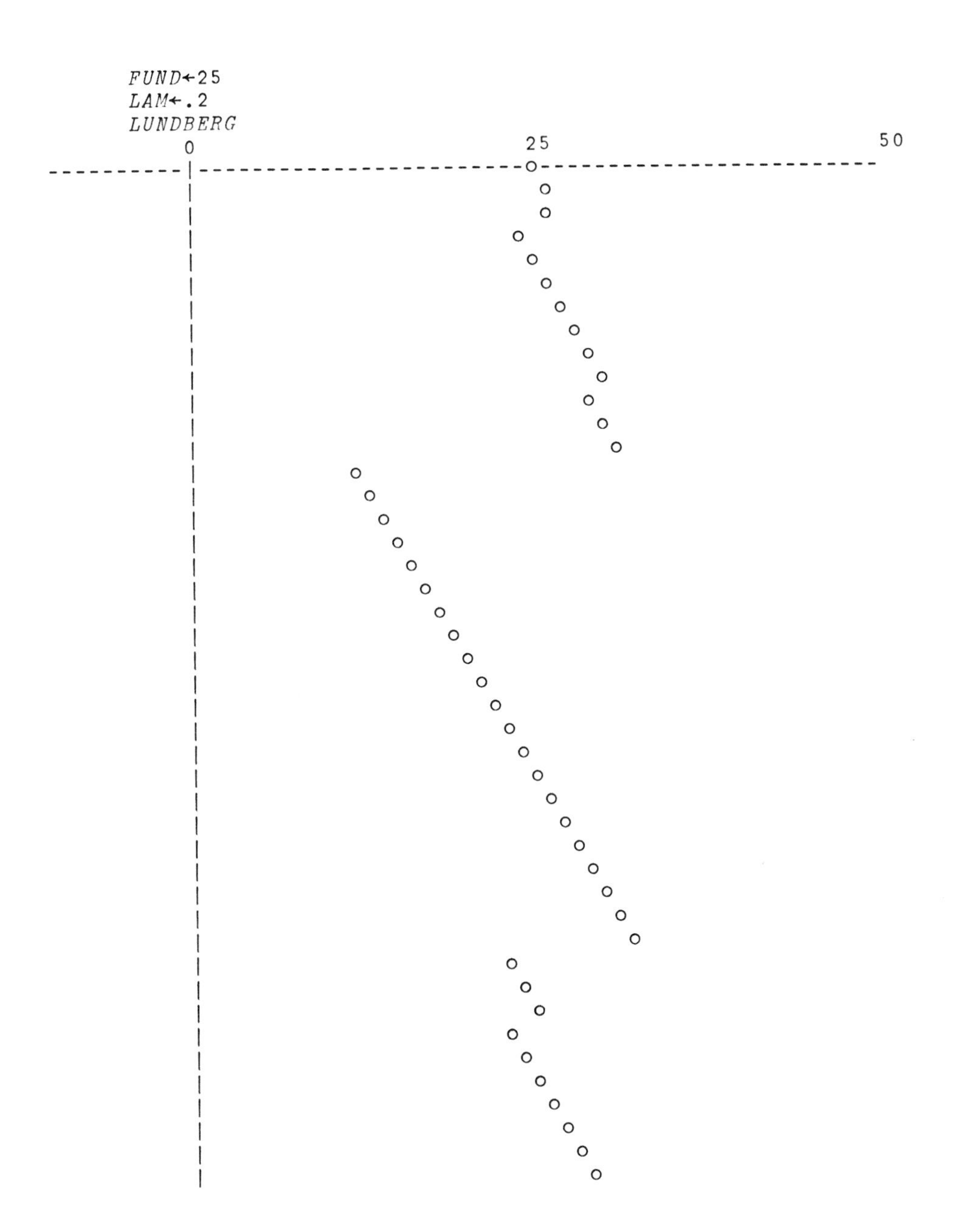

Figure 3.3

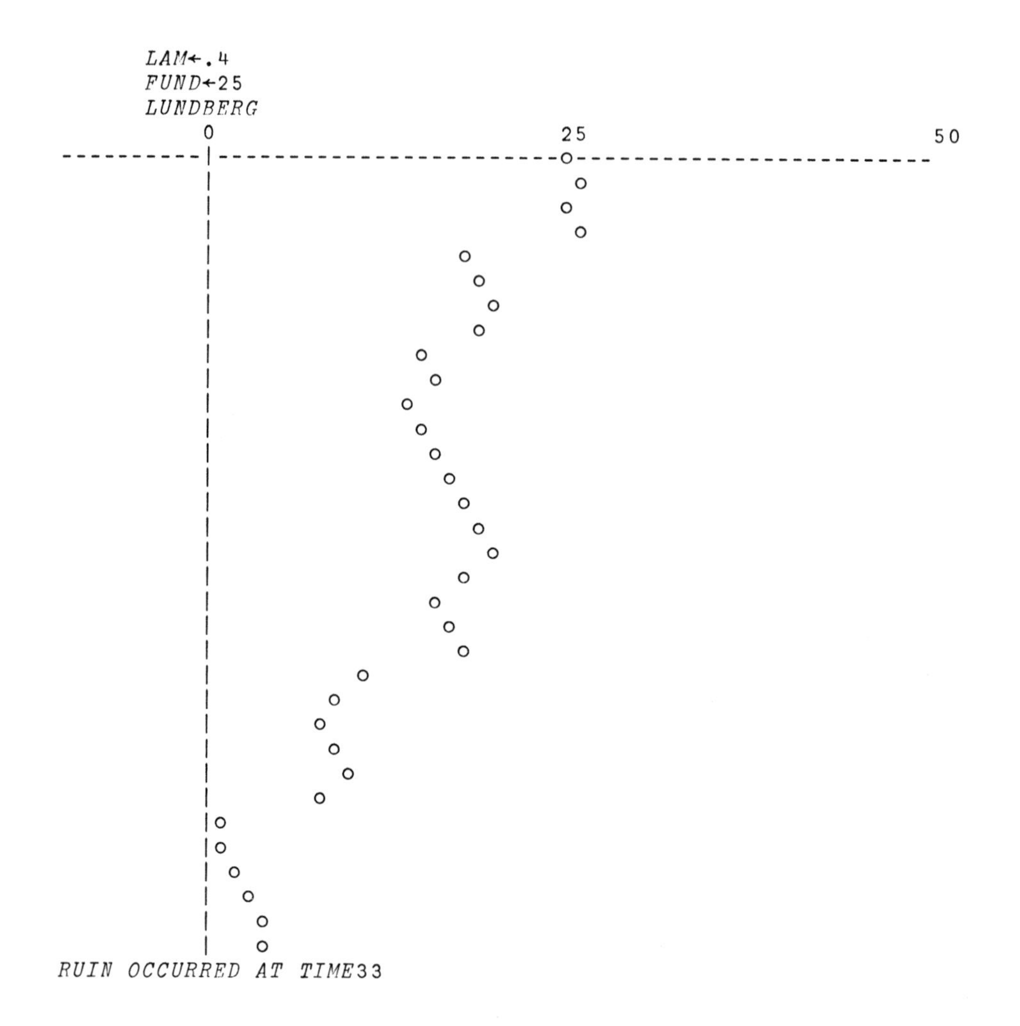

Figure 3.4

```
    ∇TAIL[⎕]∇
      ∇ TAIL
[1]     X←(¯3+2×?2)×(0.0001×?10000)*-A
      ∇
```

Figure 3.5

```
    ∇CONTPL[⎕]∇
      ∇ CONTPL;M;T;N
[1]     T←2
[2]     N←30
[3]     M←Nρ0
[4]     TAIL
[5]     M[1]←X
[6]    CONTL:TAIL
[7]     M[T]←(÷T)×X+M[T-1]×T-1
[8]     T←T+1
[9]     →(T≤N)/CONTL
[10]    M
[11]    INITIALIZE
[12]    YMAX←⎕
[13]    YMIN←⎕
[14]    DY←0.1×(YMAX-YMIN)
[15]    OFFSET 0,3,YMIN,DY
[16]    PLOT 0,YMIN,13
[17]    PLOT 0,YMAX,12
[18]    PLOT 0,YMIN,13
[19]    PLOT 30,YMIN,12
[20]    TTY
[21]    T←⍳30
[22]    PLOT T[1],M[1],13
[23]    N←2
[24]   LOOP:PLOT T[N],M[N],12
[25]    N←N+1
[26]    →(N≤30)/LOOP
[27]    PLOT T[1],M[1],13
[28]    TTY
[29]    →
      ∇
```

Figure 3.6

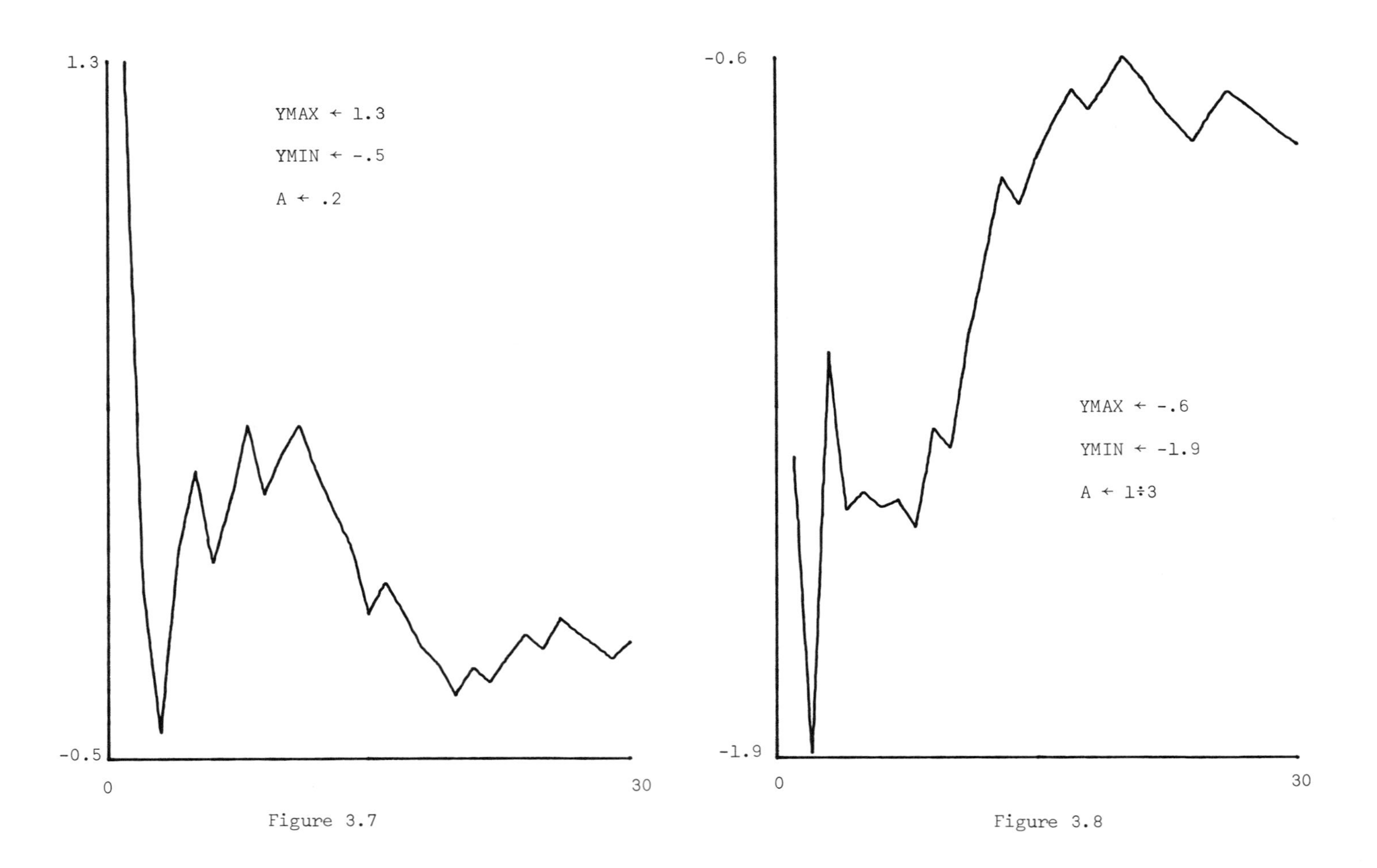

Figure 3.7

Figure 3.8

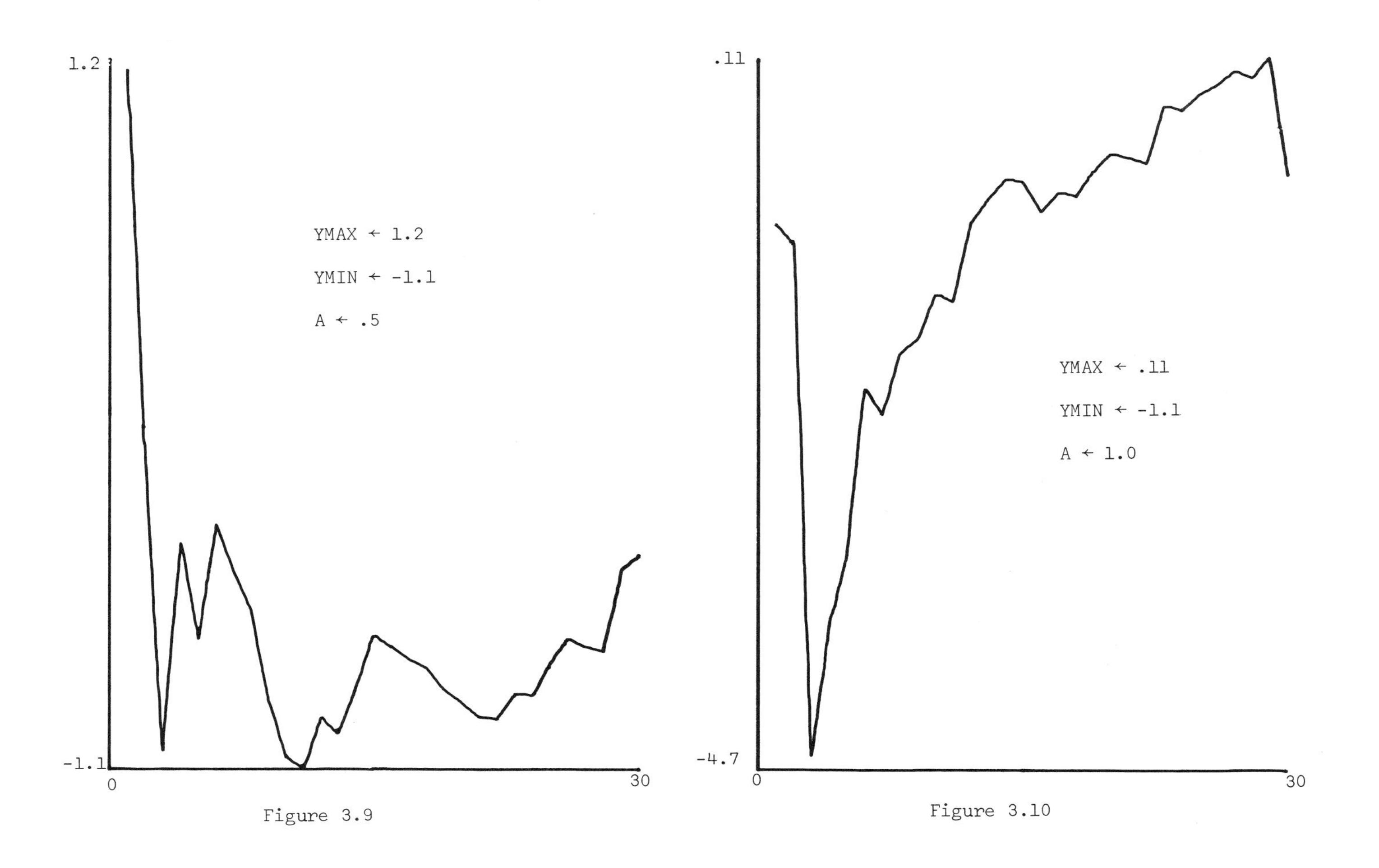

Figure 3.9

Figure 3.10

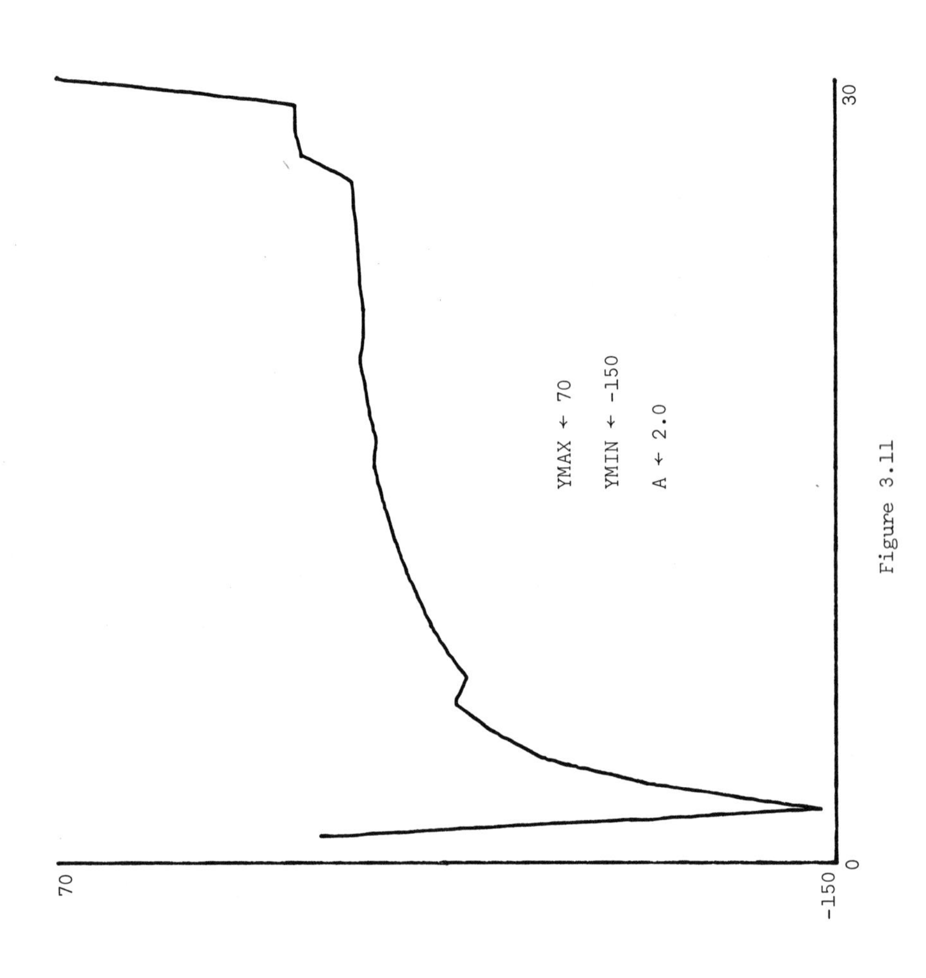

Figure 3.11

CHAPTER 4:

STOCHASTIC PROCESSES

4.1 General Properties.

We have until now been concerned mainly with sequences of independent and identically distributed stochastic variables. When we turn to situations in which we believe that either the variables are dependent upon each other or their probability distributions change with time, or both, the mathematical tools take on a quite different appearance. The relevant tool for studying such situations is the theory of random functions, that is, the theory of stochastic processes.

The formal definition of a stochastic process is simple: it is a family of indexed stochastic variables $\{x_t;\ t\varepsilon T\}$. The subscript "t" can here be of quite general type and range over some set T of values. We shall usually take T to be the set of integers, or a subset of it. This definition may be misleading because of its generality; the theory of stochastic processes becomes useful only when we specialize the definition in a more precise manner.

How do we describe the probability properties of a stochastic process x_t? A minimal, and seldom sufficient, description is obtained by giving the (first-order marginal) distribution functions

$$F_t(x) = P\{x_t \leq x\} \tag{4.1.1}$$

Knowing F_t we can calculate other quantities that will help in clarifying the structure of x_t. One such quantity is the mean value function

$$m(t) = Ex_t = \int_{-\infty}^{\infty} x\, F_t(dx) \tag{4.1.2}$$

(if the integral exists).

Knowing F_t alone does not help us in understanding the stochastic dependence between, say, x_s and x_t. This can be done by specifying the (second-order marginal) distribution functions

$$F_{s,t}(x_1,x_2) = P\{x_s \leq x_1;\ x_t \leq x_2\} \tag{4.1.3}$$

An important derived quantity is the covariance function

(4.1.4) $$r(s,t) = \mathrm{Cov}(x_s,x_t) = \int_{-\infty}^{\infty}\int_{-\infty}^{\infty} x_1 x_2 \, F_{s,t}(dx_1,dx_2) - m(s)\, m(t)$$

and the corresponding correlation function

(4.1.5) $$(s,t) = \frac{r(s,t)}{\sqrt{r(s,s)\, r(t,t)}}$$

We can of course go ahead and specify higher-order marginal distributions of order 3, 4 and so on. We shall for the moment, however, be satisfied with those of the first and second order and thus discuss second-order theory of stochastic processes.

It will be illuminating at this point to take a look at data generated on the computer, but first without knowing what the underlying program is. The curve in figure 4.2 of Appendix 4 is supposed to represent (simulated) temperature data over a period of 5 years of monthly averages at one location. Here $t = 1$ means January of the first year, $t = 2$ means February of that year, and so on. The data were obtained by the program TEMP (figure 4.1).

Looking at this realization of the stochastic process, we get some idea of the seasonal variation over the year. This periodic fluctuation is of course hidden by random changes for the individual months. To eliminate these random changes at least to some extent, we shall take 10 independent realizations $x_t^{(i)}$ $(i = 1,2,\ldots,10)$ of the same stochastic process and form the estimate m_t^* (APL program AVERAGE, fig. 4.3):

(4.1.6) $$m_t^* = \frac{1}{10}\sum_{i=1}^{10} x_t^{(i)}$$

of m_t; m_t^* is plotted in figure 4.4. It illustrates the effect of the law of large numbers (large here meaning 10) and gives a good idea of the mean value function: the time average (4.1.6) has got closer to the mean value function which is a theoretical concept, not an empirical one.

What we have done here is to form an approximation to the ensemble average (4.1.2) by using (4.1.6) and repeated independent realizations. Could we not have tried to use a time average (program TIMEAVERAGE, figure 4.5) from one realization $x_1,x_2,\ldots,x_{60}$ by

(4.1.7) $$x_t = \frac{1}{5}(x_t + x_{t+12} + x_{t+24} + x_{t+36} + x_{t+48})\ ?$$

Doing this we get the plot in figure 4.6, which also clearly brings out the seasonal variation over the year. Notice, however, that (4.1.7) is based on the assumption that the 5 sections of $x_1, x_2, \ldots, x_{60}$, consisting of one year each, have the same stochastic properties (at least as far as the mean value function is concerned). If we suspect that a secular variation, a trend, is present, we would have to be more careful about averaging over time.

The model generating x_t was in this case simply

$$(4.1.8) \qquad x_t = \text{const} + a \cdot \cos \frac{2\pi(t-7)}{12} + n_t$$

where n_t was white noise; that is, n_t were independent and identically distributed stochastic variables (in the present case n_t had, in particular, a uniform distribution over the interval (0,3)).

Note that we can speak of two types of dependence for stochastic processes:

(i) dependence of x's upon time

(ii) dependence between x's

In the model (4.1.8) we had only type (i) dependence. This implies that $r(s,t) = \rho(s,t) = 0$ if $s \neq t$. Let us now look at dependence of type (ii), which is rather more difficult to handle. We shall illustrate it by an example.

4.2 An Investment Example.

Consider an investment company where the incoming stream of capital to be invested is described by a Poisson process (see chapter 5) with intensity λ per unit time and where the size of a particular contribution has a distribution function F. The durations have a distribution function G. The interest rate is denoted by δ and we will be using a continuous-time model so that the value of \$1 after t time units is $\$e^{\delta t}$.

Assume that we start with total capital zero and study how the buildup occurs. The program INVEST (figure 4.7 of Appendix 4) simulates this and we present two graphs of output. In figure 4.8 λ is small; in figure 4.9 it is larger. In both cases the behavior seems to settle down to statistical equilibrium, slowly in the first case and more rapidly in the second. We can derive the mean value function m(t) as follows.

The total investment x_t at time t gets contributions from intervals (s,s+ds) with $0 < s < t$. The expected value of x_t is made up additively from these contributions and gives us the expression (4.2.2) by the following reasoning. An investment made in the time interval $(s, s+ds)$ will have the value $e^{\delta(t-s)}$ at time t if it is still there. The probability of this being the case is the same as the probability that the duration is at least equal to t-s:

$$(4.2.1) \qquad P(\text{duration} > t-s) = 1 - G(t-s)$$

If m is the mean value of the investment at the time when it is made, we get

$$(4.2.2) \qquad m(t) = \lambda m \int_0^t e^{\delta(t-s)} [1 - G(t-s)] \, ds$$

When t tends to infinity we have

$$(4.2.3) \qquad \lim_{t\to\infty} m(t) = \lambda m \int_0^\infty e^{\delta s} [L - G(s)] \, ds$$

if the integral is finite.

By a similar but more complicated argument we can find the covariance function $r(s,t)$. It can be shown that if s and $t \to \infty$ with $t-s = h > 0$, the limiting covariance is

$$(4.2.4) \qquad r(h) = (m^2 + \sigma^2) \lambda \int_0^\infty e^{(2s+h)\delta} [1 - F(s+h)] \, ds$$

if the expression for $r(0)$ is finite; here σ^2 stands for the variance of F.

This behavior is typical of many cases encountered in applied probability: there is first a transient stage where development occurs, then a stationary stage characterized by statistical equilibrium. The two stages are of course not separated by a clear dividing line. The transient is usually difficult to treat analytically, but for the stationary stage we have some well-developed mathematical techniques available; this leads us to the concept of a stationary stochastic process.

4.3 Stationary Stochastic Processes.

We shall say that a stochastic process x_t, with marginal distribution functions $F_{t_1,t_2,\ldots,t_n}$, is stationary if

$$(4.3.1) \qquad F_{t_1+h,t_2+h,\ldots,t_n+h} \equiv F_{t_1,t_2,\ldots,t_n}$$

This implies in particular that F_t does not depend on t so that m(t) is just some constant, say μ, and that $F_{s,t}$ depends only upon the time difference t-s; thus the covariance function r(s,t) can be written as a function of a single argument t-s, viz. r(t-s).

Let us assume that the mean m = 0; if not, we can always subtract the constant m from all the observations x_t (assuming, of course, that m is known; otherwise see chapter 8). Introduce the covariance matrix

$$(4.3.2) \qquad R = \{r_{ij};\ i,j = 1,2,\ldots,n\}$$

where elements are constant along diagonals i-j = h. Such a matrix is called a Toeplitz matrix and plays an important role in the theory of stationary stochastic processes; see ref. 20.

The matrix (4.3.2) is non-negative definite; that is, for any vector z ≠ 0 in R^n we have

$$(4.3.3) \qquad z^*Rz = \sum_{i,j=1}^{n} z_i z_j r_{i-j} \geq 0$$

since

$$(4.3.4) \qquad z^*Rz = E(\sum_{\nu=1}^{n} z_\nu x_\nu)^2 \geq 0$$

A consequence of this property is the theorem of Herglotz that asserts that there is a uniquely determined bounded measure F, the spectral distribution function, such that

$$(4.3.5) \qquad r_t = 2\int_0^{\pi} \cos t\lambda\ F(d\lambda);\quad t = 0,\pm1,\pm2,\ \ldots$$

The program HERGLOTZ effecting this Fourier inversion will be given in chapter 8. We shall often deal with the case where F is absolutely continuous with a density f = F', so that

$$(4.3.6) \qquad r_t = 2\int_0^{2\pi} \cos t\lambda\ f(\lambda)\ d\lambda;\quad t = 0,\pm1,\pm2,\ \ldots$$

Of course, when F is represented in the machine it will often be practical to approximate it discretely, so that f is then really a vector with a finite number of components. In the special case of a constant spectral density (white noise) r_t vanishes except for t = 0, so that R is simply a diagonal matrix. Otherwise the behavior of the non-diagonal elements characterizes the dependence between the x's. The importance of the stationary process consists in its providing a very simple way of modelling dependence, but it is of course limited to situations where we believe that statistical equilibrium is a good approximation to reality.

The second-order moments do not specify the probability distributions uniquely. One possible and certainly the most frequently used specification is the assumption that the joint probability distribution is normal with a frequency function

$$p(x) = \frac{1}{(2\pi)^{n/2}\sqrt{\det R}} \exp(-\tfrac{1}{2}\, x'R^{-1}x) \tag{4.3.7}$$

where R has been assumed non-singular (a fact which actually follows from the representation (4.3.6)). We note the appearance of the inverse R^{-1} which is typical of many problems in stationary stochastic processes. If the inversion cannot be done in closed form it has to be done numerically, and it is then useful to take advantage of the Toeplitz character of the covariance matrix to save space and computing time (see the APL function TINV, figure 4.10 of Appendix 4).

<u>Assignment</u>. Let x_t be a stochastic process with a known covariance sequence $r_0, r_1, r_2, \ldots$. In order to predict x_{n+1} when $x_1, x_2, \ldots, x_n$ have been observed, we use a linear function

$$x^*_{n+1} = \sum_{\nu=1}^{n} c_\nu x_\nu \tag{4.3.8}$$

where the c's are chosen so that the mean square error is minimized:

$$MSE = E(x^*_{n+1} - x_{n+1})^2 = \min$$

This gives us, for $k = 1,2,\ldots,n$

$$\frac{\partial MSE}{\partial c_k} = 2E\Big(\sum_{\nu=1}^{n} c_\nu x_\nu - x_{n+1}\Big)x_k = 0 \tag{4.3.9}$$

or, in matrix notation,

$$Rc = \rho, \quad c = R^{-1}\rho \tag{4.3.10}$$

where c is the column vector $(c_1, c_1, \ldots, c_n)$ and ρ the column vector $(r_1, r_2, \ldots, r_n)$. This gives

(4.3.11) $$MSE_{min} = r_0 - c^* R^{-1} \rho$$

First, choose a covariance sequence and compute MSE_{min} for different values of n; what do you observe?

Secondly, assume that the r's are not known but that we use our observations $x_1, x_2, \ldots, x_n$ to estimate them by

(4.3.12) $$r_\nu^* = \frac{1}{n} \sum_{\nu=1}^{n-h} x_\nu x_{\nu+h}; \quad h = 0,1,2,\ldots,\ell$$

To use the last ℓ observed values, we replace R by the $\ell \times \ell$ matrix R^* with entries as in (4.3.12), and similarly for ρ. Study how the resulting prediction error varies with ℓ and try to explain the somewhat paradoxical result in an intuitive way.

4.4 Markov Chains.

The two main classes of stochastic processes are the stationary ones and the Markov processes. We now turn to the simplest type of Markov process, the Markov chain. We shall examine a particular case which brings out some of the computational problems arising in the study of Markov chains.

Consider a paged computing system; that is, one in which the program to be executed does not reside totally in high-speed store during execution but is brought in "page" by "page" as instructions in a certain page are referenced. Enumerate the pages available in some secondary memory by $1,2,3,\ldots,n$ and assume that core (also organized in pages of the same size) can contain c pages, where c is considerably less than n. Assume that the pages being referenced successively are $n(1),n(2),n(3),\ldots$. If a referenced page is not already in core it has to be pulled in and some other page pushed out. The decision which page to push out is determined by some rule, the paging algorithm, whose purpose it is to reduce the paging rate, that is the probability that a page fault occurs at a given time.

If a page fault occurred at a certain time t, and if we knew what the future of the sequence $n(t+1),n(t+2),n(t+3),\ldots$ was, we could reduce the paging to a minimum (assuming here that the resulting overhead computing is acceptable). This is, however, not the case, so we have to use probabilistic considerations.

Let us try the following model. The sequence $\{n(t);\ t = 1,2,3,\ldots\}$ forms a Markov process with transition probabilities

(4.4.1) $$p_{in} = P\{n(t+1) = j | n(t)=i\}$$

forming an $n \times n$ matrix P. For a given initial load in core and with the knowledge of P we can, at least in principle, get the expected paging rate for a particular paging algorithm. In this way we can compare competing algorithms and should be able to make a rational choice.

There are, however, several obstacles in the way. An obvious one is that the problem may be analytically untractable; we have therefore to take recourse to simulation. This may not be a serious drawback.

We meet a difficulty of a different order of magnitude when we try to specify P. Little is known empirically about the statistical properties of $\{n(t)\}$ and even if we had access to good data there would be nothing making it plausible that P will not change in time, between users, and between installations. This forces us to look for algorithms that are not based on a priori knowledge of P, although we can of course gain some insight into the problem by studying how well we could do if we postulated the values of P.

We shall below assume the following block structure for P, which seems motivated by the modular form of many programs and which has also received some empirical support. The values $1,2,3,\ldots,n$ are separated into blocks

(4.4.2) $$\begin{cases} B_1: & 1,2,\ldots,b_2-1 \\ B_2: & b_2,b_2+1,\ldots,b_3-1 \\ B_3: & b_3,b_3+1,\ldots,b_4-1 \\ \cdot & \cdot \quad \cdot \quad \cdot \\ B_r: & b_r,b_r+1,\ldots,n \end{cases}$$

A stochastic $r \times r$ matrix

(4.4.3) $$\Pi = \{\Pi_{k\ell};\ k,\ell = 1,2,\ldots,r\}$$

describes the frequency of transitions between blocks of pages. Between blocks we assume a uniform distribution over pages and inside blocks a cyclic behavior, so that if i and j belong to different blocks k and ℓ

(4.4.4) $$p_{ij} = \Pi_{k\ell} \cdot (1/\text{number of pages in block } \ell)$$

and if i and j belong to the same block k

$$(4.4.5) \qquad p_{ij} = \begin{cases} \Pi_{kk} & \text{if } j = 1 + i \text{ modulo the length of block } k \\ 0 & \text{otherwise} \end{cases}$$

Consider the following paging algorithms.

RANDOM: The page to be pushed out is selected at random from the ones in core. Computationally simple.

FIFO (first in, first out): The "oldest" page in core is pushed out. This is also easy to compute.

LIFO (last in, first out): The last page is always pushed out (or a simplification of this).

PROB1: The page j in core for which p_j = min is pushed out, where p_j stands for the equilibrium probability.

PROB2: Push out the page j for which p_{ij} min, where i is the page being brought into core.

PROB3: Push out the page j for which m_{ij} = max, where m_{ij} is expected time until j is referenced for the first time after i.

The last three algorithms may require a bit more overhead computing. Unfortunately, they also require knowledge of P. A more adaptive approach is via

EXP1: Keep track of the frequencies f_j with which the various pages are being referenced and push out j where f_j = min.

To find the m_{ij} for the algorithm PROB3 we can reason as follows. Assume that we are referencing page i and look one step further ahead. Then (assuming that the mean values exist)

$$(4.4.6) \qquad m_{ij} = \sum_{k \neq j} p_{ik}(m_{kj} + 1) + p_{ij} = 1 + \sum_{k \neq j} p_{ik} m_{kj}$$

so that the column vector $m_j = (m_{1j}, m_{2j}, \ldots, m_{nj})$ satisfies

$$(4.4.7) \qquad (I - A_j) m_j = e$$

e being the column vector (1,1,...,1) and A_i the P matrix with the elements in the jth column replaced by zeroes. Hence

$$(4.4.8) \qquad m_j = (I - A_j)^{-1} e$$

Assignments. 1. Simulate the behavior of at least two paging algorithms for given n (say 20) and Π. Note that the approach to equilibrium may take a long time if the initial load of pages in core is unfavorable for some algorithms.

Do this for different values of c between 1 and n and plot the observed paging rule against c.

2. A program consists of n blocks $B_1, B_2, B_3, \ldots, B_n$ of sizes $s_1, s_2, s_3, \ldots, s_n$; the sizes are expressed in terms of some multiple of a byte. The blocks call each other in an order that depends on the input data; some calls $B_i \to B_j$ occur often, others are less frequent. Assume that these transition probabilities p_{ij} are (at least approximately) known and that the calls can be regarded as a stationary Markov chain.

We want to combine the calls into an arbitrary number r of modules $M_1, M_2, \ldots, M_r$, where the only restriction is that the size of any module must be less than some given constant c (bytes); c may be thought of as the page size. We could write

$$M_1 = B_{i_1}, B_{i_2}, \ldots$$

$$M_2 = B_{j_1}, B_{j_2}, \ldots$$

$$\cdot \quad \cdot \quad \cdot \quad \cdot \quad \cdot$$

with the restriction that

$$s_{i_1} + s_{i_2} + \ldots \leq c$$

$$s_{j_1} + s_{j_2} + \ldots \leq c$$

$$\cdot \quad \cdot \quad \cdot \quad \cdot \quad \cdot$$

The partition $B_1 \cup B_2 \cup B_3 \cup \ldots = M_1 \cup M_2 \cup M_3 \cup \ldots$ should, of course, be a partition into mutually exclusive modules and be exhaustive. This rearrangement has the aim that any module should call any other module as seldom as possible. More precisely, for the complementary probability Π of calling the same module we want

$$\Pi = \sum_k \sum_{(i,j) \varepsilon M_k} \Pi_{ij} = \max$$

where Π_{ij} is the absolute probability of the transition $B_i \to B_j$. Hence, if

$$\text{TRANS} = \{p_{ij}\}$$

we first have to find the equilibrium distribution $\Pi_1, \Pi_2, \ldots, \Pi_n$ of the Markov chains satisfying

$$\Pi_i = \sum_k \Pi_k \, p_{ki}$$

or, in matrix form,

$$EQ = EQ \cdot TRANS$$

where $EQ = \Pi_1, \Pi_2, \ldots, \Pi_n$. Computationally it may be better to get EQ by forming $TRANS^2$, $TRANS^4$, $TRANS^8$, etc., and take EQ = first row of the limit. This will also give some idea of how long it takes to reach statistical equilibrium. Once we have obtained EQ, by one method or another, we get $\Pi_{ij} = \Pi_i p_{ij}$.

Try this for the matrix:

$$TRANS = \begin{bmatrix} .4 & .1 & .1 & .3 & .1 \\ .1 & .5 & .3 & .05 & .05 \\ .05 & .4 & .5 & .05 & 0 \\ .4 & .1 & .1 & .35 & .05 \\ .05 & .4 & .1 & .05 & .4 \end{bmatrix}$$

and with $c = 8$, $s_1 = 2$, $s_2 = 2$, $s_3 = 6$, $s_4 = 5$, $s_5 = 3$. Try different partitions including the improper one with $B_1 = M_1$, $B_2 = M_2$, $B_3 = M_3$, $B_4 = M_4$, $B_5 = M_5$.

The APL programs for this are given in Appendix 4 as figures 4.11 and 4.12.

```
∇TEMP[⎕]∇
    ∇ TEMP
[1]    X←15+(10×2○2×○(¯7+⍳60)÷12)+0.03×?60⍴100
[2]    M←(60 2)⍴0
[3]    M[;1]←⍳60
[4]    M[;2]←X
    ∇
```

Figure 4.1

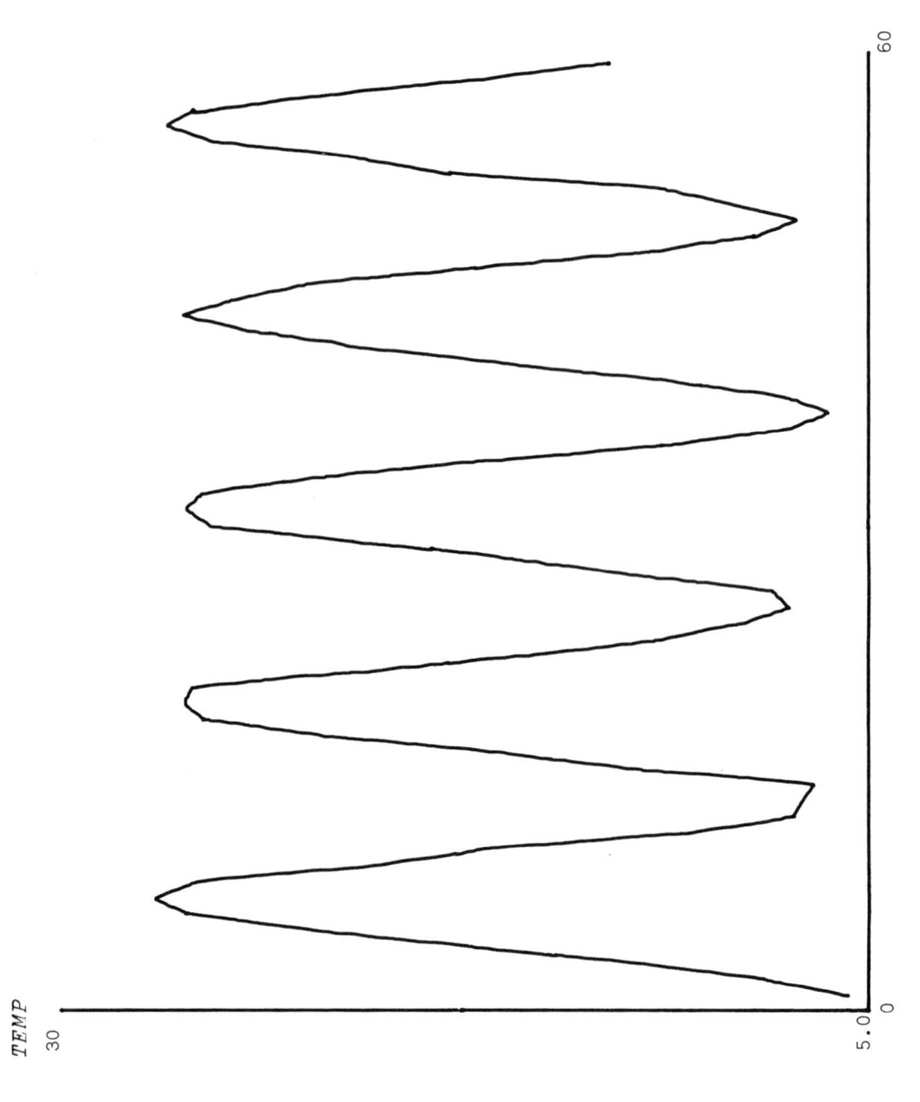

Figure 4.2

```
        ∇AVERAGE[⎕]∇
     ∇ AVERAGE
[1]    Y←0
[2]    X←0
[3]    N←0
[4]    N←N+1
[5]    TEMP
[6]    Y←Y+X
[7]    →(→(N≤10)/4
[8]    M←(÷10)×Y
     ∇
```

Figure 4.3

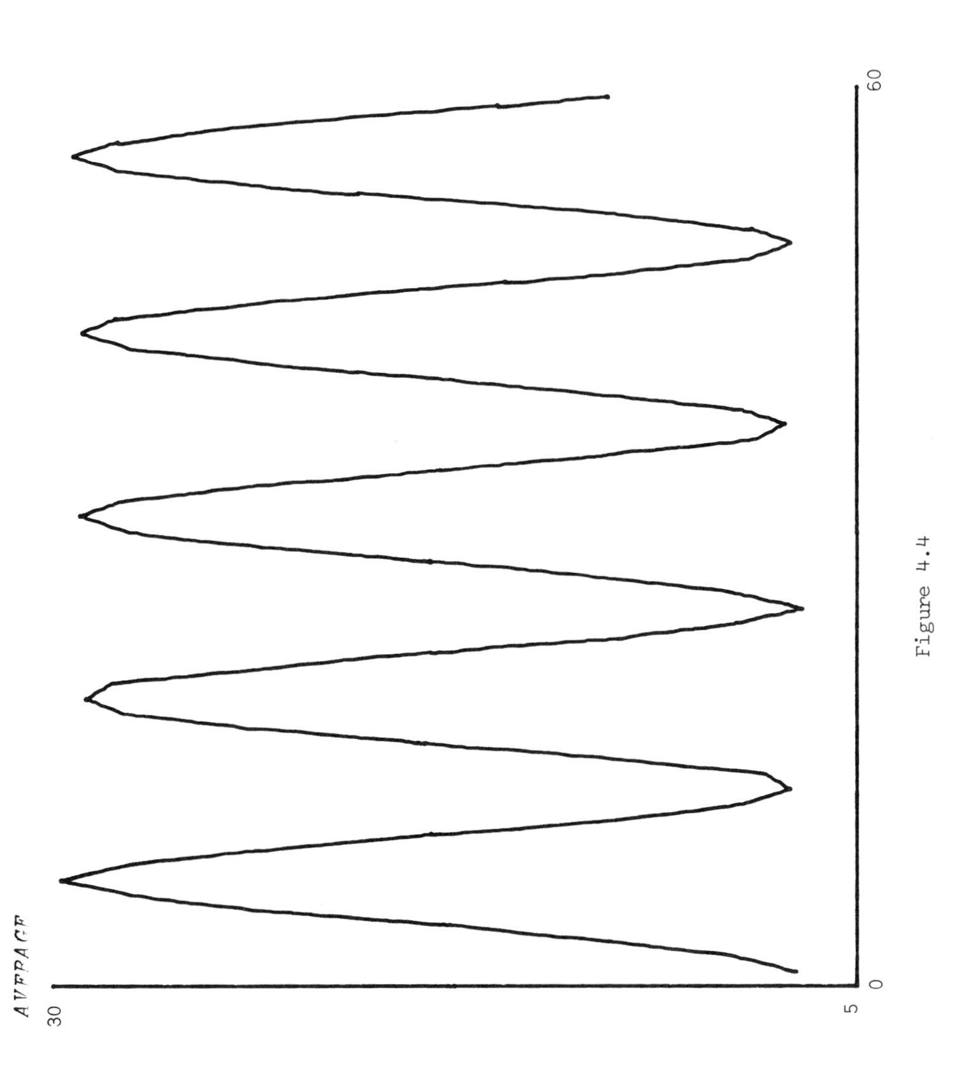

Figure 4.4

```
 ∇TIMEAVERAGE[⎕]∇
    ∇ TIMEAVERAGE
[1]   TEMP
[2]   Y←(÷5)×X[⍳12]+X[12+⍳12]+X[24+⍳12]+X[36+⍳12]+X[48+⍳12]
    ∇
```

Figure 4.5

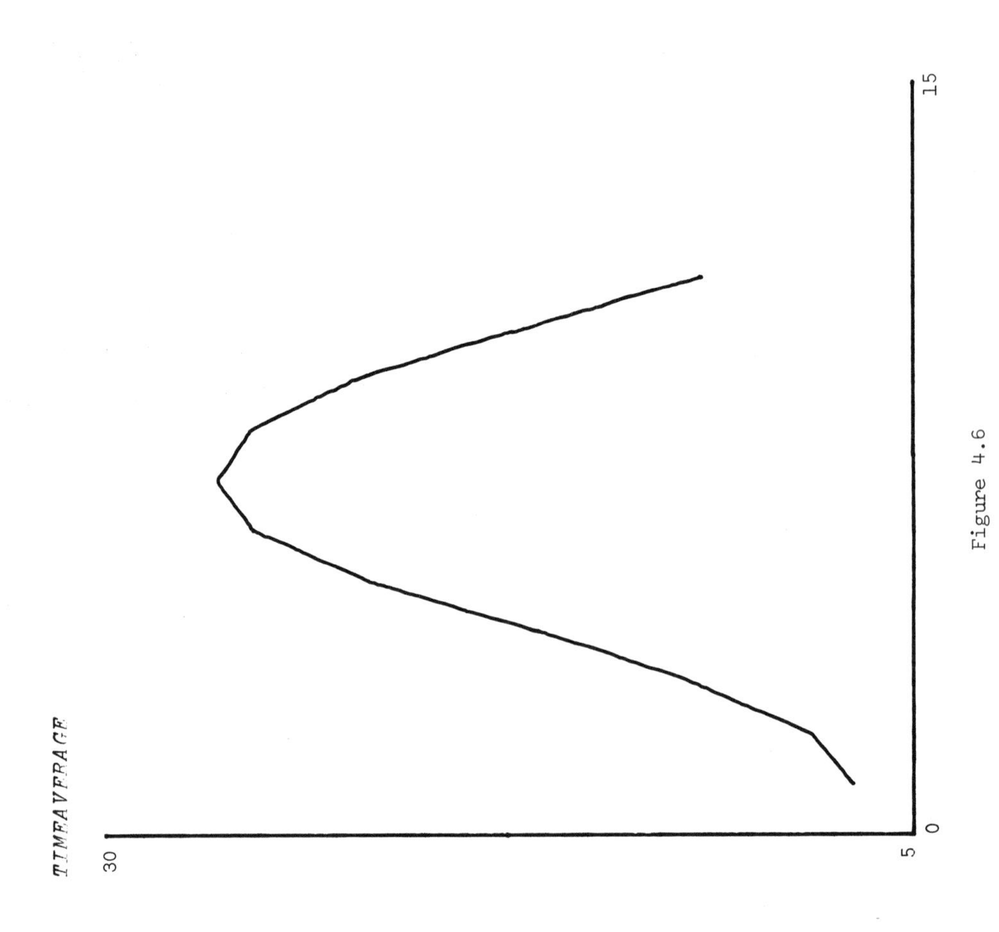

Figure 4.6

```
      ∇INVEST[⎕]∇
    ∇ INVEST
[1]    STATE←DUR←⍳0
[2]    T←1
[3]    TOTAL←RUN⍴0
[4]    R←⍴F
[5]    L←⍴MU
[6]    TRI←(⍳R)∘.≥⍳R
[7]    CUMF←TRI+.×F
[8]   BANKLO:→((⍴,STATE)>0)/BANKL1
[9]    TOTAL[T]←0
[10]   →BANKL2
[11]  BANKL1:TOTAL[T]←+/STATE
[12]   PDROP←MU[L⌊DUR]
[13]   DROP←(0.001×?(⍴STATE)⍴1000)≤PDROP
[14]   DUR←(1-DROP)/DUR
[15]   STATE←(1-DROP)/STATE
[16]  BANKL2:E←P≥0.001×?1000
[17]   DUR←DUR+1
[18]   →(1-E)/BANKL4
[19]   STATE←STATE,1++/CUMF<0.001×?1000
[20]   DUR←DUR,1
[21]  BANKL4:T←T+1
[22]   STATE←STATE×1+INTEREST×TSTEP
[23]   →(T≤RUN]/BANKLO
    ∇
```

Figure 4.7

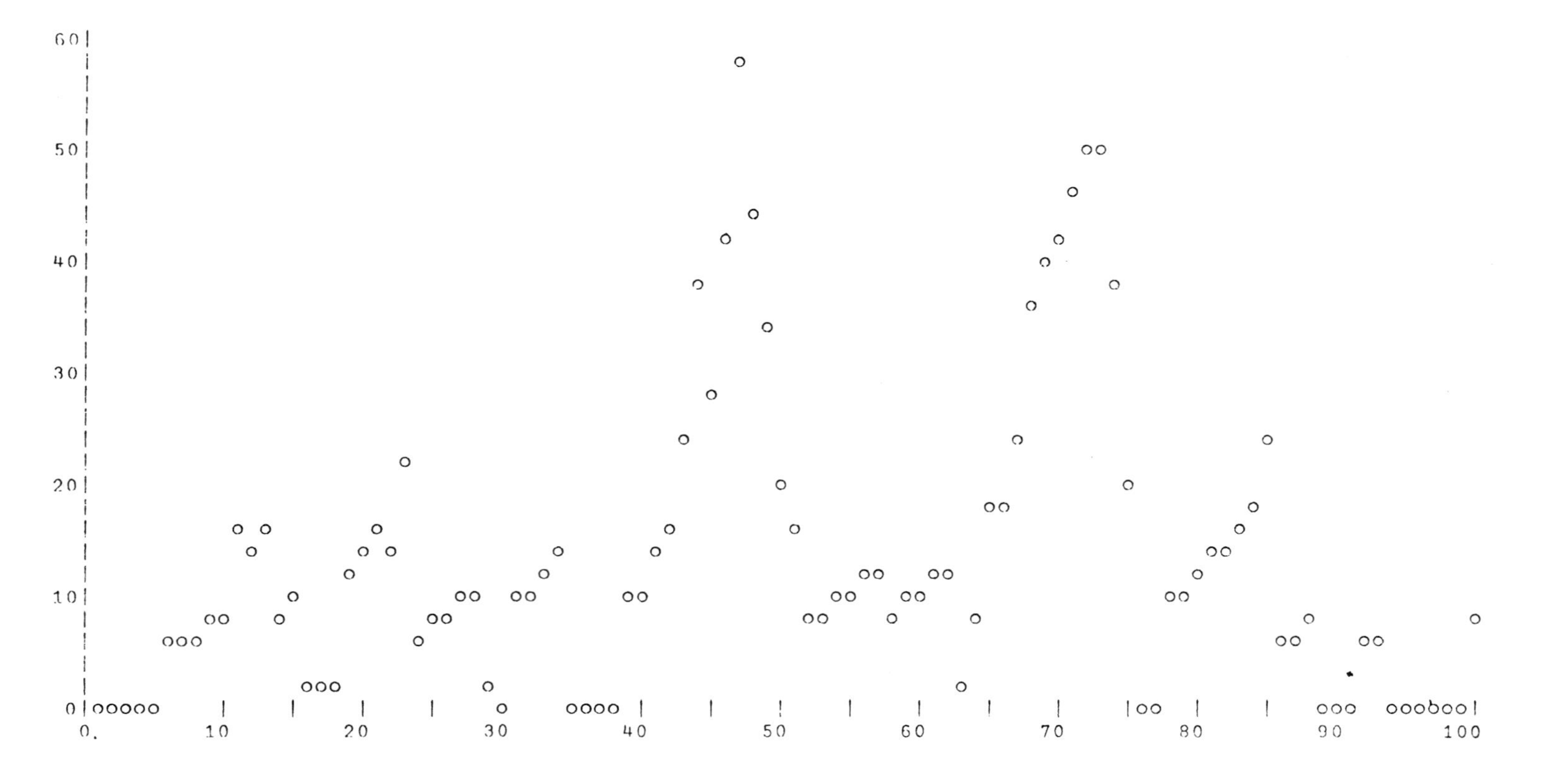

Figure 4.8

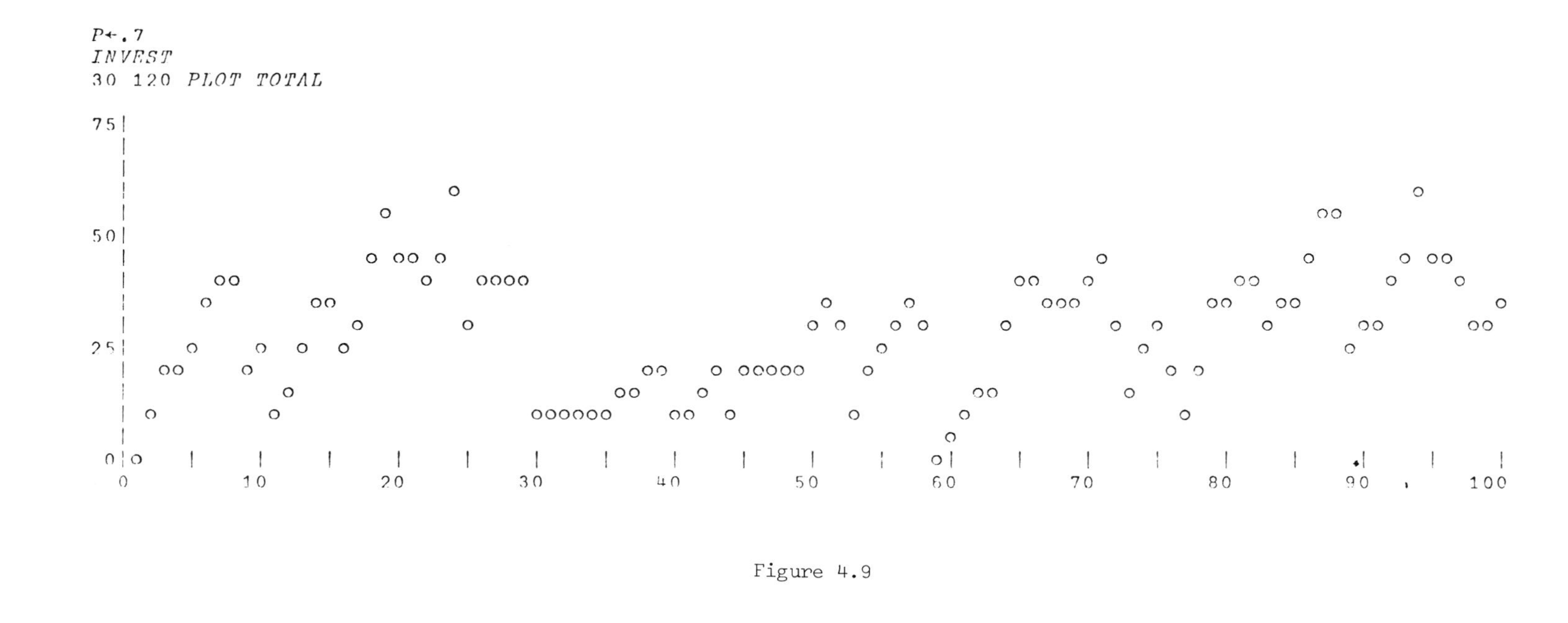
P←.7
INVEST
30 120 PLOT TOTAL
75
50
25
0
0
10
20
30
40
50
60
70
80
90
100

Figure 4.9

```
∇TINV[⎕]∇
    ∇ Z←TINV R;LA;E;N;I;G;M
[1]    →12×⍳1=⍴,R
[2]    DET←LA←1-E×E←-,1↑R←1↓R÷L←1↑R←(⍴R)[1]↑,R
[3]    →7×⍳2≥N←(⍴R)+¯1+I←2
[4]    E←(E,0)+((⌽E),1)×(÷LA)×G←-R[I]++/E×⌽,(I-1)↑R
[5]    DET←DET×LA←LA-G×G÷LA
[6]    →4×⍳N>I←I+1
[7]    Z←(1,E)÷LA
[8]    →11×⍳0=M←⌊(N-1)÷2×I←1
[9]    Z←Z,Z[I+1+N×¯1+⍳I],(((I-1)↓(-I+1)↓(-N)↑Z)+(÷LA)×(I-1)↓(-I)↓(E[I]×E)-E[N-I]×⌽E),⌽,Z[(-I)+N×⍳I]
[10]   →9×⍳M≥I←I+1
[11]   →0,,Z←(N,N)⍴(÷L)×Z,((2|N)×N)↓⌽Z
[12]   →0,,Z←(1 1)⍴÷DET←R
    ∇
```

Figure 4.10

```
        ∇EQUIL[⎕]∇
    ∇ EQUIL TRANS
[1]   I←1
[2]  EL:TRANS←TRANS+.×TRANS
[3]   E←⌈/(⌈/[1] TRANS)-⌊/[1] TRANS
[4]   I←I+1
[5]   →(E>ERROR)/EL
[6]   EQ←TRANS[1;]
    ∇
```

Figure 4.11

```
        ∇REARR[⎕]∇
    ∇ REARR TRANS
[1]   N←(⍴TRANS)[1]
[2]   I←1
[3]   B←(N,N)⍴0
[4]  RL1:'TYPE BLOCK NO. ';I
[5]   V←,⎕
[6]   B[I;⍳⍴V]←V
[7]   I←I+1
[8]   →((+/+/B[⍳I-1;]>0)<N)/RL1
[9]   NBLOCKS←I-1
[10]  →((⌈/+/(L[B+B=0]×B>0))≤PAGESIZE)/RL2
[11]  'IMPOSSIBLE REARRANGEMENT'
[12]  →0
[13] RL2:TRANS←(⍉(N,N)⍴EQ)×TRANS
[14]  I←1
[15]  SUM←0
[16] RL3:V←(B[I;]>0)/B[I;]
[17]  SUM←SUM++/+/TRANS[V;V]
[18]  I←I+1
[19]  →(I≤NBLOCKS)/RL3
[20]  'PAGING RATE WILL BE ';1-SUM
    ∇
```

Figure 4.12

CHAPTER 5:

PARTICULAR STOCHASTIC PROCESSES

5.1 A Growth Model.

We now turn to special types of stochastic processes met in applications. Consider the following problem involving a growth model.

A population is described as follows. The number $n(t)$ of individuals at a certain time t can fluctuate between the values 0 and N. We shall use a birth and death process such that if $n(t) = n$ then the conditional distribution at $t+h$ is given by

$$(5.1.1) \qquad \begin{cases} P[n(t+h) = n-1] = \mu_n h + o(h) \\ P[n(t+h) = n] = 1 - \mu_n h - \lambda_n h + o(h) \\ P[n(t+h) = n+1] = \lambda_n h + o(h) \end{cases}$$

for $0 < n < N$, and with obvious modifications if $n = 0$ or $n = N$. This means that μ_n plays the role of the death intensity and λ_n of the birth intensity. Introducing the probability functions

$$(5.1.2) \qquad p_n(t) = P[n(t) = n]$$

we get

$$(5.1.3) \quad p_n(t+h) = p_{n-1}(t)\,\lambda_{n-1}h + p_n(t)[1-\lambda_n h-\mu_n h] + p_{n+1}(t)\,\mu_{n+1}h + o(h)$$

and modified relations for $n = 0$ and $n = N$. This leads to the classical differential equation for a birth and death process

$$(5.1.4) \quad \begin{cases} dp_0(t)/dt = -\lambda_0 p_0 + \mu_1 p_1 \\ dp_n(t)/dt = \lambda_{n-1}p_{n-1} - (\lambda_n+p_n)p_n + \mu_{n+1}p_{n+1}; \qquad 0 < n < N \\ dp_N(t)/dt = \lambda_{N-1}p_{N-1} - \mu_N p_N \end{cases}$$

If the population is of size i at time $t = 0$ we have to solve the differential equations in (5.1.4) together with the initial conditions

$$(5.1.5) \qquad p_n(0) = \begin{cases} 1 \text{ if } n = i \\ 0 \text{ otherwise} \end{cases}$$

We shall be interested in what happens as t grows and shall look at the particular case of constant mortality $\mu_n = \mu$ and a birthrate intensity of the form

$$\lambda_n = a - b(\tfrac{n}{N})^2 \tag{5.1.6}$$

It is obvious that $\{p_n(t)\}$ will approach an equilibrium distribution $\pi_0,\pi_1,\ldots,\pi_N$ satisfying

$$\begin{aligned} 0 &= -\lambda_0\pi_0 + \mu_1\pi_1 \\ 0 &= \lambda_{n-1}\pi_{n-1} - (\lambda_n+\mu_n)\pi_n + \mu_{n+1}p_{n+1} \quad ; \quad 0 < n < N \\ 0 &= \lambda_{N-1}\pi_{N-1} - \mu_N p_N \end{aligned} \tag{5.1.7}$$

This is just (5.1.4) with the time derivatives equal to zero.

It is less easy to get some idea of the speed of convergence: how fast does the transient die out? It is true that (5.1.4) looks simple enough, since it is just a finite system of first-order differential equations with constant coefficients, so that the solutions are combinations of exponential functions (with possible degeneracies). This would make an analytic treatment possible. In a practical situation a numerical integration of (5.1.4) may, however, be preferable and represent a wiser allocation of human resources. We also rule out the simulation approach here since we can, with a computing effort of about the same size, get a more definitive answer from the numerical integration. We shall later on meet a case where the opposite holds true.

Let us now make a comment on the scheme of numerical integration. We shall start from (5.1.3), leaving out the o(h) term. When we approximate a continuous-time model with a discrete one we have to make sure that the solution possesses the qualitative features asked of it _a priori_. In the present case $\{p_n(t)\}$ should be a probability distribution, so that the features needed are 1) non-negative values and 2) values summing to one. We can guarantee this here since in going from t to t+h we form convex combinations (if $h\lambda_n+h\mu_n \le 1$) so that positivity is preserved. Summing the equation (5.1.3) we get

$$\sum_{n=0}^{N} p_n(t + h) = \sum_{n=0}^{N} p_n(t) \tag{5.1.8}$$

so that the total sum is preserved: if we started with the value one in the initial distribution $\{p_n(0)\}$, $\{p_n(t)\}$ would be a probability distribution for all t.

As an illustration we display the graphs representing the transient behavior in figures 5.2 through 5.5, of Appendix 5. Look also at the program GROW given in figure 5.1. We have used N = 20, μ = 1, a = 2 and b = 1, i = 4.

5.2 The Random Phase Model.

The following stochastic process is sometimes a realistic model. It is defined for integral values of t, and its value x_t is assumed to be made up additively from individual harmonic contributions of the type $A\cos(\lambda t+\phi)$. Here A is the amplitude, λ the period and ϕ the phase. We get

$$x_t = \sum_{\nu=1}^{p} A_\nu \cos(\lambda_\nu t + \phi_\nu) \tag{5.2.1}$$

To specify the probabilistic properties of the process we shall ask that

(5.2.2)
(i) A_ν's and λ_ν's are given numbers
(ii) ϕ_ν's are given stochastic variables, independent and $R(0,2\pi)$

The mean value function is

$$m_t = \sum_{\nu=1}^{p} A_\nu E\cos(\lambda_\nu t + \phi_\nu) = \sum_{\nu=1}^{p} A_\nu \frac{1}{2\pi}\int_{\phi=0}^{2\pi} \cos(\lambda_\nu t + \phi)d\phi = 0 \tag{5.2.3}$$

and the covariance function is obtained as

$$\begin{aligned} r(s,t) = Ex_s x_t &= \sum_{\nu=1}^{p} A_\nu^2 \frac{1}{2\pi}\int_{\phi=0}^{2\pi} \cos(\lambda_\nu t + \phi)\cos(\lambda_\nu s + \phi)d\phi \\ &= \frac{1}{2}\sum_{\nu=1}^{p} A_\nu^2 \int_{\phi=0}^{2\pi} \{\cos^2(\lambda_\nu(t-s)) + \cos(\lambda_\nu t+\lambda_\nu s+2\phi)\, ds \\ &= \frac{1}{2}\sum_{\nu=1}^{p} A_\nu^2 \cos^2 \lambda_\nu(t-s) \end{aligned} \tag{5.2.4}$$

Hence, in terms of (4.3.5), we have a discrete spectral distribution with jumps at the frequencies λ_ν and of size $A_\nu^2/4 = F_\nu$.

Note that r(s,t) depends only upon the time difference t-s. It can also be shown that (4.3.1) holds so that the stochastic process is actually stationary. The reader is advised to try to prove that this is so.

This model is, for obvious reasons, called the random phase mode; we shall have occasion to return to it in chapter 8.

Assignment. For a given vector F of length M with positive components, write a program to generate a stochastic process for $t = 1,2,3,\ldots,N$ according to (5.2.1) where we have for the frequencies $\lambda_\nu = \pi\nu/m$ the spectral mass F_ν = the νth component above. Use the result to study the mean value and covariance functions. Take N = 20, M = 30. Note the following. When we form the time average

$$m^* = \frac{1}{N}\sum_{t=1}^{N} x_t \tag{5.2.5}$$

we cannot be quite sure that the law of large numbers will operate to our advantage: the x's are not independent (also, of course, the sample size N is rather small here). Does the estimate m^* behave in a stable manner when you iterate the procedure in (5.2.5)?

5.3 Renewal Processes.

An important extension of the Poisson process is the renewal process, which is defined as follows.

On the real line events occur at certain time points $\ldots t_{-1}, t_0, t_1, t_2, \ldots$ defined through the random mechanism

(5.3.1)

(i) all time differences $s_\nu = t_{\nu+1} - t_\nu$ are distributed according to one and the same probability distribution F.

(ii) all time differences s_ν are stochastically independent of each other.

We also need some kind of critical condition to start the process. Here we shall do this by assuming that $t_0 = 0$ and only consider positive values of ν. Another possibility would be to assume that statistical equilibrium has been reached, so that the probability of finding one event in an interval $(t,t+h)$ will depend only upon h, not upon t.

A sort of mean value function m(t) is introduced as

$$m(t) = E\{\text{number of events in } (0,t)\} \tag{5.3.2}$$

and we shall now get an analytical expression for it.

(5.3.3) $$p_n = P\{\text{exactly } n \text{ events in } (0,t)\} = P\{t_n \le t \text{ and } t_{n+1} > t\}$$

Notice that t_n can be regarded as the sum of n independent stochastic variables, each with the distribution function F. Hence

(5.3.4) $$p_n = P\{t_n \le t\} - P\{t_{n+1} \le t\} = F^{n*}(t) - F^{(n+1)*}(t)$$

Therefore

(5.3.5) $$m(t) = \sum_{n=1}^{\infty} n\, p_n = F(t) - F^{2*}(t) + 2F^{2*}(t) - 2F^{3*}(t) + 3F^{3*}(t) - 3F^{4*}(t) \dots\dots\dots$$

which, assuming that the series converges, can be rearranged as

(5.3.6) $$m(t) = \sum_{n=1}^{\infty} F^{n*}(t)$$

Also note the following relation, obtained by convolving (5.3.6) with F:

(5.3.7) $$m * F = \sum_{2}^{\infty} F^{n*} = m - F$$

which is the integral equation

(5.3.8) $$m(t) = F(t) + \int_0^t F(t-s)\, m(ds)$$

a classical relation known as the <u>renewal equation</u>. Numerically it may be easier to get m by solving (5.3.8) since the system of linear equations which is its discrete counterpart can be solved directly with no explicit matrix inversion.

The program RENEWAL (see figure 5.6 of Appendix 5) does this. It seems plausible that as t becomes large m(t) should grow linearly in the asymptotic sense. Indeed, the law of large numbers tells us that, if the mean value μ of F is finite, then $t_n/n \to \mu$. This in turn makes the above statement seem reasonable. To see whether we can get some experimental support for this conjecture, we run RENEWAL with the probabilities .2, .1, .3, .4 at the values t = 1, 2, 3, 4. We get the result shown in figure 5.7. It demonstrates a convergent behavior for $m(t)/t$, although a rather slow one, to $1/\mu$, which is in this case .3448.

For an analytic discussion of the renewal problem, see ref. 16, Chapter XI.

```
        ∇GROW[⎕]∇
     ∇ GROW RUN
[1]    T←1
[2]    N←20
[3]    P←(N+1)⍴0
[4]    INITIAL←4
[5]    P[INITIAL]←1
[6]    H←1
[7]    MU←(N+1)⍴1
[8]    LAM←2-((⍳22-1)÷20)*2
[9]    LAM[1]←2
[10]   A←H×LAM
[11]   C←H×MU
[12]   B←(N+1)⍴1-A-C
[13] GL:V1←0,¯1↓A×P
[14]   V2←P×B
[15]   V3←(1↓C×P),0
[16]   P←V1+V2+V3
[17]   T←T+1
[18]   →(T≤RUN)/GL
[19]   30 PLOT P
     ∇
```

Figure 5.1

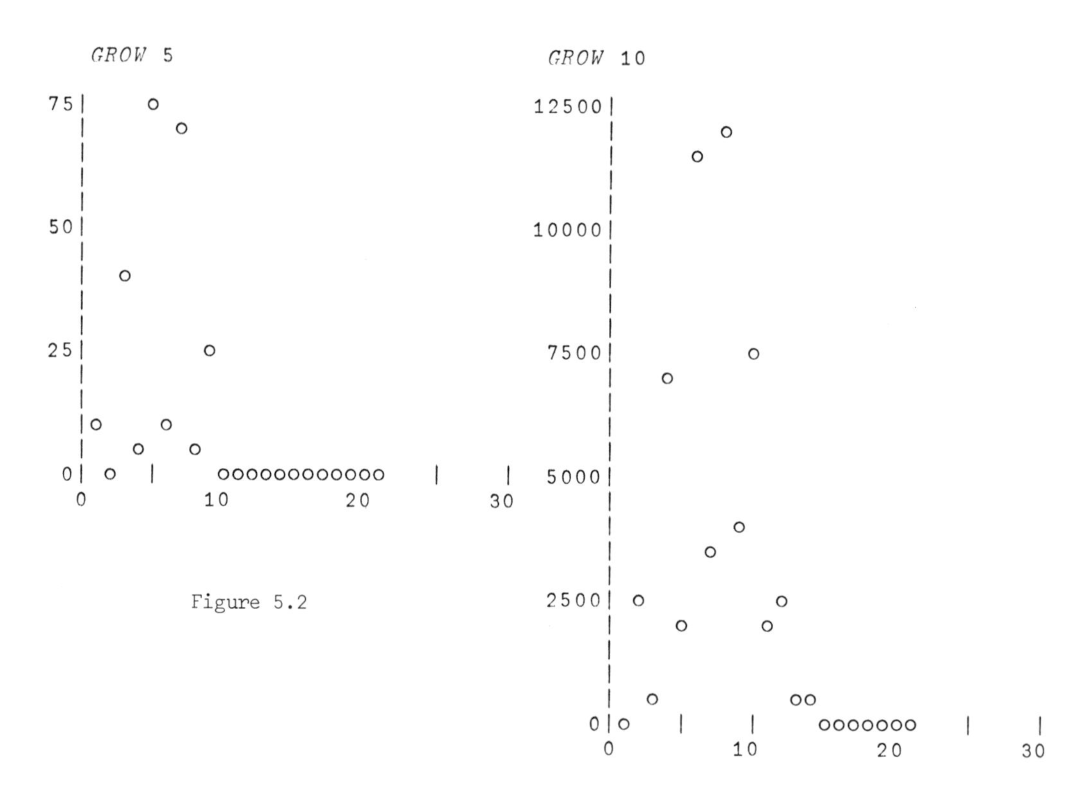

Figure 5.2

Figure 5.3

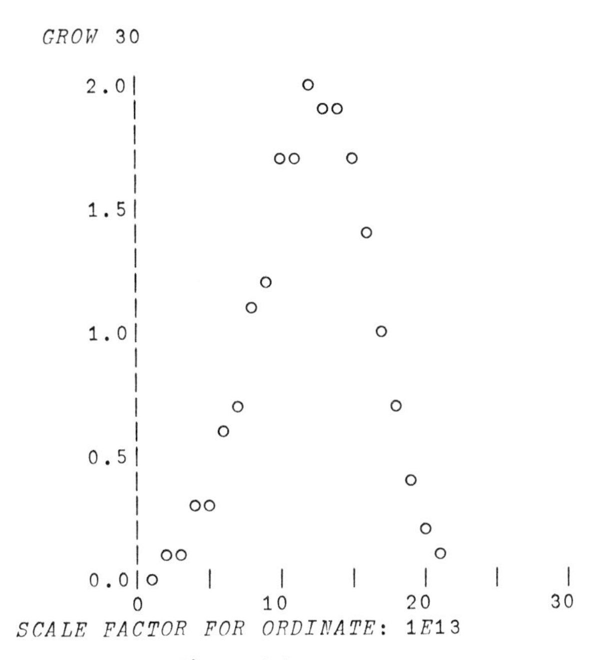

Figure 5.4

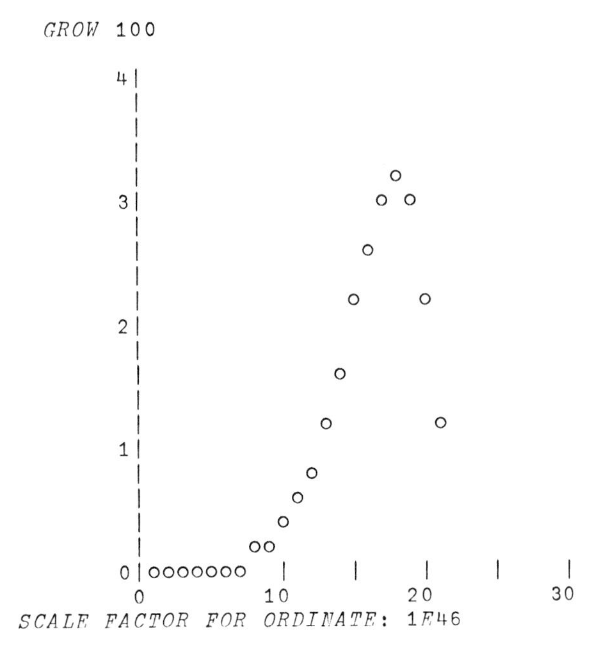

Figure 5.5

```
      ∇RENEWAL[⎕]∇
    ∇ N RENEWAL F
[1]   F←F,(N-ρF)ρ1
[2]   M←Nρ0
[3]   M[1]←F[1]
[4]   K←2
[5]  RL:M[K]←+/(F[⍳K]-0,F[⍳K-2],0)×⌽1,M[⍳K-1]
[6]   K←K+1
[7]   →(K≤N)/RL
[8]   ⍉(3,N)ρ(⍳N),M,M÷⍳N
    ∇
```

Figure 5.6

```
40 RENEWAL .2 .3 .6 1
```

1	0.2	0.2000
2	0.34	0.1700
3	0.688	0.2293
4	1.232	0.3079
5	1.497	0.2994
6	1.765	0.2942
7	2.147	0.3068
8	2.548	0.3185
9	2.853	0.3170
10	3.176	0.3176
11	3.544	0.3221
12	3.901	0.3251
13	4.228	0.3253
14	4.569	0.3264
15	4.924	0.3283
16	5.271	0.3294
17	5.609	0.3299
18	5.954	0.3308
19	6.303	0.3317
20	6.647	0.3323
21	6.989	0.3328
22	7.335	0.3334
23	7.681	0.3340
24	8.025	0.3344
25	8.369	0.3348
26	8.715	0.3352
27	9.06	0.3355
28	9.404	0.3359
29	9.749	0.3362
30	10.09	0.3365
31	10.44	0.3367
32	10.78	0.3370
33	11.13	0.3372
34	11.47	0.3374
35	11.82	0.3377
36	12.16	0.3379
37	12.51	0.3380
38	12.85	0.3382
39	13.2	0.3384
40	13.54	0.3386

Figure 5.7

CHAPTER 6:

DECISION PROBLEMS

6.1 Generalities.

One formulation of the general decision problem starts from a set of possible stochastic models specifying the resulting probability distributions P_α on some sample space X. Here α is a general parameter in some parameter space A. Both X and A can, of course, be high-dimensional vector spaces or function spaces.

Using an observation $x \in X$ we want to select a decision d from the set D of available decisions, and we want to do it in such a manner that the expected cost is made small. This assumes a cost function $c(d,\alpha)$, which is the cost associated with the decision d if P_α is the true probability distribution; the parameter is assumed to be unknown to the decision-maker. In other words, we compute the risk associated with a particular decision function

$$\delta : X \to D \tag{6.1.1}$$

as

$$R(\delta,\alpha) = \int_X c(\delta(x),\alpha)\, P_\alpha(dx) \tag{6.1.2}$$

The problem is then reduced to finding that (or those) δ that give

$$\min_\delta R(\delta,\alpha) \tag{6.1.3}$$

if such functions δ exist.

But here we meet what has been called the uniformity problem in decision theory. It may be that (6.1.3) cannot be solved because of its dependence upon α. To illustrate this, assume that we have access to only two decision functions, δ_1 and δ_2, and that their risk functions look as in figure 1. Here it is obvious that none of the δ's achieve the desired minimum uniformly in α, and this is not to be considered an atypical case, but on the contrary the usual one.

Many approaches have been suggested to avoid this difficulty, and we shall mention two here.

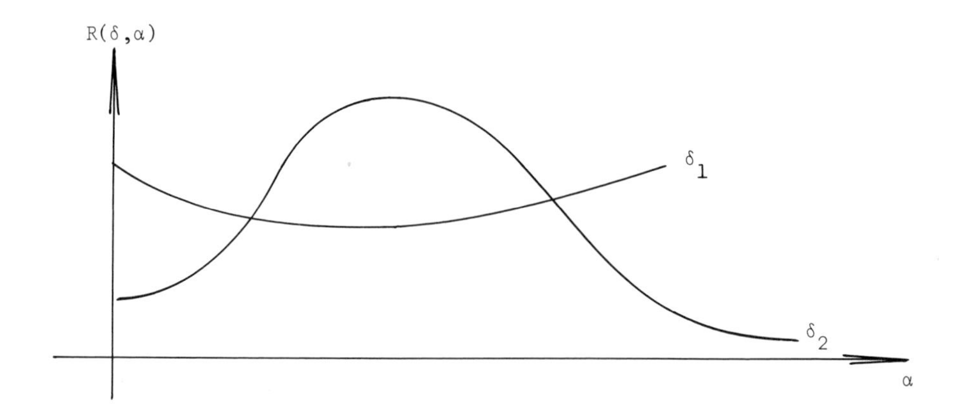

Figure 1

The first is the Bayes approach, in which one assumes that α can be treated as a stochastic variable with some (known) *a priori* probability distribution $\Pi(\alpha)$. We can then ask for a Bayes solution δ_B such that

$$E_{\Pi}\, R(\delta_B,\alpha) = \int_A R(\delta_B,\alpha)\, \Pi(d\alpha) = \min_{\alpha} \tag{6.1.4}$$

It does not, sometimes, even make sense to speak of a probability distribution over the state space A; at other times it may make sense but the distribution may be unknown. The computational difficulties associated with the Bayes approach have in recent years come to be realized. These, rather than philosophical difficulties, are the main stumbling block, and they are due to the frequent high dimensionality of the state space. It should be remembered that the state space is the *a posteriori* probability *function*, and is thus a function space rather than a low-dimensional vector space. The quest for higher computational efficiency - for "fast Bayes methods" - deserves more study.

Another approach is to appeal to the *mini-max* criterion and ask for a solution δ_M solving the extremum problem

$$\max_{\alpha} R(\delta,\alpha) = \min_{\alpha} \tag{6.1.5}$$

This is quite conservative: we look at the worst possible value of α for a given decision function δ and then try to make this maximum as small as possible.

In order to make the preceding discussion more concrete we shall consider the following problem, belonging to the area called stochastic approximation.

6.2 A Stochastic Approximation Problem.

A stochastic process $y(a)$, where a takes discrete and equidistant values in $(0,1)$, is made up of two parts, a systematic component $m(a)$ and a noise term $\varepsilon(a)$:

$$y(a) = m(a) + \varepsilon(a) \tag{6.2.1}$$

Having access to p observations $y(a_1), y(a_2), \ldots, y(a_p)$, we want to estimate $m(a)$ by some linear estimate $m^*(a)$ so that

$$E \int_0^1 [m(a) - m^*(a)]^2 \, da = \min \tag{6.2.2}$$

The term $\varepsilon(a)$ represents white noise, with $E\varepsilon(a) \equiv 0$. This makes it reasonable to use as a cost function $c = -I$ where

$$I = \int_0^1 [m(a) - m^*(a)]^2 \, da \tag{6.2.3}$$

We now add the assumption that m is generated by some stochastic mechanism, so that we can deal with $m(a)$ as a stochastic process. To be more specific, the *a priori* distribution of m is specified by the relation $m(a) = k \cdot a(1-a) + n(a)$, where $n(a)$ has mean zero and is formed as a moving average

$$n(a) = \sigma \sum_\nu w_\nu \, \eta_{a+\nu} \tag{6.2.3a}$$

in white noise η. Both η and ε are assumed to have jointly normal probability distributions, η independent of ε, $E\eta = 0$, $E\eta^2 = 1$.

It is possible to show that, whatever the probability distributions are, to minimize (6.2.2) we should use the conditional expected value (or regression)

$$m^*(a) = E[m(a) \mid y(a_1), y(a_2), \ldots, y(a_p)] \tag{6.2.4}$$

But since regression functions for normal distributions are known to be linear, (6.2.4) is equivalent to finding the best linear approximation (regression) of the form

$$m^*(a) = ca(1-a) + \sum c_\nu \, [n(a_\nu) + \varepsilon(a_\nu)] \tag{6.2.5}$$

so that

(6.2.6) $$Q = E(n(a) - \sum c_\nu[n(a_\nu) + \varepsilon(a_\nu)]^2 = \min_{c_\nu}$$

Expanding Q, we get

$$Q = R(0) + \sum c_\nu c_\mu \; [R(a_\nu - a_\mu) + \delta_{\nu\mu}\sigma^2] - 2 \sum c_\nu \; R(a_\nu - a)$$

where R is the covariance function of the n-process, so that to achieve the minimum in (6.2.6) the column vector $c = (c_1, c_2, \ldots, c_p)$ should be

(6.2.7) $$c = (R + \sigma^2 I)^{-1} \rho_a$$

where ρ_a stands for the column vector $R(a-a_1), R(a-a_2), \ldots, R(a-a_p)$. The resulting form of Q is then

(6.2.8) $$Q_{min} = R(0) - \rho_a^T (R + \sigma^2 I)^{-1} \rho_a$$

and m^* is given by (6.2.5).

This solution has been coded (see the programs in Appendix 6) and can conveniently be used together with a graphical representation routine.

A similar problem of considerable interest is obtained when we have access to experimental data as above and want to use them in order to plan a further experiment designed for the purpose of locating the maximum of m(a). The reader is advised to experiment with different strategies: locate the a's in such a way that as much information as possible is obtained concerning that portion of the m-function that is of interest, say the region where m takes its largest values. A set of programs is given in Appendix 6, together with descriptions of them and the output of a run. Most of the programming for this section was done by Mr. R. Lepping, a graduate student at Brown University.

6.3 An Insurance Game.

Two insurance companies compete for portfolios of policies. In each portfolio claims are generated as follows during one time period (see also section 3.1). The number of claims has a Poisson distribution with parameter α, the size of a claim has a Pareto frequency function

$$(6.3.1) \qquad f(x) = \begin{cases} \text{const} \cdot x^{-\beta}, & x > \gamma \\ 0, & x \leq \gamma \end{cases}$$

and we assume independence between all these stochastic variables.

Now we take a Bayesian point of view and assume some probabilistic mechanism that gives rise to the values α, β, γ and appeal to this a priori probability mechanism once for each portfolio.

For each time-period each player (company) bids on each portfolio using his previous experience, except in the first period when no experience is available. The bid has the form of a premium. The lowest bid wins the portfolio.

Now we see what happens during the following period. If the total claim for one portfolio exceeded the premium this would result in a loss for the player who won the contract, and vice versa. The other players would not be affected.

Going through several time periods, the players try to adjust their bids in order to maximize profit. At present it is intended that they do their risk-taking intuitively, but the interested reader is advised to look into algorithmic-analytic alternatives.

The game could be played by two (or more) companies, each company being made up of 5-10 persons. A referee sits at the terminal and presents each company's data to that company but not to the others.

6.4 Design of Experiments.

There are indications that interactive computing will be a potent tool for the design of experiments. To get some feeling for the way in which this may develop, we shall study a special case.

Assume that we can observe a stochastic variable $y(x)$ depending upon a design parameter x, $0 \leq x \leq 1$, such that

$$(6.4.1) \qquad y(x) = f(x) + n(x)$$

where the stochastic term, the statistical noise $n(x)$, will be assumed to be $N(0,\sigma^2)$, known σ, independent observations and with some deterministic component that is not completely known to us. Our resources allow us to take a sample of size S but we can allocate them to different values of x.

Assume that we use the r+1 values $x = 0,1/r,2/r,\ldots,(r-1)/r,1$. At each of these values of x we allocate n observations, so that $n = [S/(r+1]$. Let us assume that f(x) is a polynomial of order p but we are not sure that the term of order p is necessary. Hence we want to test the value of p.

First, it is clear that an argument based on the concept of sufficient statistics would show that the data are equivalent to having just one observation at each x but with σ^2 replaced by σ^2/n. It is convenient to use the orthogonal polynomials up to order p over the set of x's; i.e.

(6.4.2) $$u_\nu(x) = c_{\nu 0} + c_{\nu 1}x + c_{\nu 2}x^2 + \ldots + c_{\nu\nu}x^\nu, \quad \nu = 0,1,2,\ldots,p$$

and

$$\sum_x u_\nu(x)\, u_\mu(x) = \begin{cases} 1 \text{ if } \nu = \mu \\ 0 \text{ otherwise} \end{cases}$$

To get these polynomials we use the classical Gram-Schmidt orthonormalization procedure. Initialize

(6.4.3) $$u_0(x) = c_{00} = 1/\sqrt{r+1}$$

and proceed recursively. If we have defined $u_0(x),u_1(x),\ldots,u_{\nu-1}(x)$ so that they satisfy (6.4.2), we can write

(6.4.4) $$p(x) = x^\nu - c_0u_0(x) - c_1u_1(x) \ldots - c_{\nu-1}u_{\nu-1}(x)$$

To make p(x) orthogonal to $u_0,u_1,\ldots,u_{\nu-1}$, we must have

(6.4.5) $$\sum_x x^\nu\, u_\nu(x) = c_\mu\,, \quad \mu = 0,1,2,\ldots,\nu-1$$

(6.4.6) $$p(x) = x^\nu - \sum_{\mu=0}^{\nu-1} \sum_z x^\nu\, u_\mu(z)\, u_\mu(x)$$

so that we can take

(6.4.7) $$u_\nu(x) = \frac{p(x)}{\sqrt{\Sigma p^2(x)}}$$

This has been coded as the APL program GRAM (see Appendix 6).

Let us write

(6.4.8) $$f(x) = f_0 + f_1 x + f_2 x^2 + \dots + f_\nu x^\nu$$
$$= g_0 u_0(x) + g_1 u_1(x) + \dots + g_\nu u_\nu(x)$$

so that a linear estimate of f_ν can be expressed as

(6.4.9) $$f^* = \sum e_x\, y(x)$$

In order that f^* be unbiased, we must have, identically for all $f_0, f_1, \dots, f_\nu$,

(6.4.10) $$Ef^* \equiv \sum e_x\, f(x) \equiv f$$

and the variance is

(6.4.11) $$\mathrm{Var}(f^*) = \frac{\sigma^2}{n} \sum e_x^2$$

From (6.4.10) it follows, using the orthogonality of the u's, that e_x must have the form

(6.4.12) $$e_x = \alpha_\nu\, u_\nu(x) + \alpha_{\nu+1}\, u_{\nu+1}(x) + \dots$$

with

(6.4.13) $$\alpha_\nu = 1/\sum_\nu x^\nu u_\nu(x)$$

and

(6.4.14) $$\mathrm{Var}(f^*) = \frac{\sigma^2}{n} [\alpha_\nu^2 + \alpha_{\nu+1}^2 + \dots]$$

The design problem is thus reduced to making the denominator in (6.4.13) as large as possible.

Let us do this numerically; see the APL program DESIGN in Appendix 6.

Assignment. Investigate, for different choices of ν and n, the efficiency of different designs.

How can the above be modified if other functions than powers x^n are used in the function f?

Note that in a case like this direct computing of the efficiency is preferable to simulation. This may not be true if the model is more complex.

6.5 A Search Problem.

Consider the following decision problem. An object is contained in one of N boxes numbered 1,2,...,N. To find out in which box it is we are allowed to ask questions of the type: is it contained in one of the L+1 adjacent boxes A,A+1,A+2,...,B? The answer we will get is, however, not wholly reliable, and is true with some probability W(L) which depends on the "length" L = B-A. If we ask for only one box, L = 0, the probability W(0) is close to one. If we ask for more boxes simultaneously we assume W(L) to decrease toward 50% (which is the least informative case). Since we know that there is exactly one object in the boxes it is realistic to assume a symmetric property for W(L): W(L) = W(N-1-L) for L = 0,1,2,...,N-1. To see this, we think of asking for A=1, B=L and compare this with asking for A=N-1-L, B=N.

When we begin our search we have only some imprecise knowledge where the object is most likely to be. We formalize this knowledge in the form of an a priori distribution $P = P_1,P_2,\ldots,P_N$. (We shall assume that all $P_i > 0$; otherwise we could just remove some boxes from consideration.) Assume that we select values A, B and get the answer that the object is inside one of the selected boxes ("answer I"). What is the a posteriori distribution $Q = Q_1,Q_2,\ldots,Q_N$?

$$(6.5.1)\qquad Q_i = P\{\text{object in } i \mid \text{answer I}\} = \frac{P\{\text{object in } i \text{ and answer I}\}}{P(\text{answer I})} = \frac{P_i \cdot P\{\text{answer I object in } i\}}{P(\text{answer I})}$$

But we find that

$$(6.5.2)\qquad P\{\text{answer I} \mid \text{object in } i\} = P\{\text{"answer I" is true}\} = W(L)$$

if $i \varepsilon [A,B]$ and

$$(6.5.3)\qquad P\{\text{answer I} \mid \text{object in } i\} = P\{\text{"answer I" is false}\} = 1 - W(L)$$

if $i \notin [A,B]$. This can also be used to compute

$$(6.5.4) \qquad P(\text{answer } I) = \sum_{1}^{N} p_i \cdot P(\text{answer } I|\text{object in } i)$$

so that we now have Q. The case when we get the answer that the object is not in one of the selected boxes ("answer 0") can be treated in a similar way.

The next step would be to look at Q and make a decision. If Q is well concentrated, say in j, we may decide to state: the object is in j. If Q is concentrated over j_1, j_2 so that most of the probability mass is in j_1 and j_2 together we may state: the object is either in j_1 or in j_2, and so on.

So far the procedure is quite straightforward, but we have left out the important question how to decide on A, B in the first place. Can knowledge of P guide us in this decision?

What is it we want to achieve? We want, after getting the answer, to be able to make as precise a statement as possible; that is, to point to as small as possible a set $j_1, j_2, \ldots$. In other words, we want to make Q as concentrated as possible. Let us measure concentration by the criterion

$$(6.5.5) \qquad \text{CRIT} = \sum_{1}^{N} Q_i^2$$

Note that concentration, in the present context, should not necessarily mean that the $j_1, j_2, \ldots$ are close together, but only that they are few in number. CRIT does this: if CRIT is large Q is concentrated.

CRIT is, however, a stochastic variable, so that we had better use its expected value ECRIT. We now compute ECRIT for different values of A, B and pick that pair AMAX, BMAX for which ECRIT is as large as possible.

The programs LENGTH, SEARCH and DECIDE have been written for the above purpose (see Appendix 6). LENGTH just specifies W(L) in a certain way.

Assignment. Use SEARCH to find the object, using your judgment only: do not use DECIDE. Go on with repeated questions until you are reasonably sure where the object is. Then stop the program and ask for the value of the variable I which is the true position of the object. Can you develop some feeling for what decision strategy you have found useful?

Now use also DECIDE for picking good values of A, B in the first trial (or question). It is advantageous to use it in one trial after another if P is replaced each time by the successively updated *a posteriori* distribution, which is called PI ("answer I") or PO ("answer 0") (see the program). To be able to point out the values of PI, PO corresponding to the choice A = AMAX, B = BMAX, a small modification is needed in the program DECIDE, inserting two statements between statements 15 and 16.

Discuss the results and suggest alternative decision procedures such as basing the choice of A, B on some other criterion, for instance the information content (entropy)

$$-\sum_{1}^{N} P_i \ln P_i \tag{6.5.6}$$

APPENDIX 6: FIGURES

To experiment with the problem discussed in section 6.2 we use the following programs.

Normally distributed stochastic variables are generated by GAUSS, sample size N, mean value PARAM[1], standard deviation PARAM[2].

```
          ∇GAUSS[⎕]∇
      ∇ RESULT←PARAM GAUSS N
[1]     RESULT←PARAM[1]+PARAM[2]×¯6+0.001×+/¯1+?(N,12)ρ1001
      ∇
```

Figure 6.1

These variables are used in RLSERIES which computes the moving average process of (6.2.3a).

```
∇RLSERIES[⎕]∇
    ∇ X←A RLSERIES LEN;AL;EL;K;SDEV;E
[1]    AL←ρA
[2]    EL←LEN+AL+5
[3]    X←LENρ0
[4]    E←(0 1) GAUSS EL
[5]    K←0
[6]   LOPP:K←K+1
[7]    X[K]←A+.×AL↑K↓E
[8]    →LOPP×ι) K≤LEN-1)
    ∇
```

Figure 6.2

To do the matrix inversion in (6.2.7) we use INVP from the public library. A more efficient program for this purpose would be TINV (figure 4.10).

```
∇INVP[⎕]∇
    ∇ Z←INVP M;I;J;K;P;S
[1]     M←⍉(1 0 +ρM)ρ(,⍉M),~J←1<P←ιI←1↑ρM
[2]     S←÷⌈/|M
[3]    L:Z←|M[ιI;1]×I↑S
[4]     K←Zι⌈/Z
[5]     M[K,1;ιρP]←M[1,K;ιρP]
[6]     S[K,1]←S[1,K]
[7]     P[K,1]←P[1,K]
[8]     P←1⌽P
[9]     S←1⌽S
[10]    M[1;]←M[1;]÷M[1;1]
[11]    M←1⌽(J,1)⊖M-(J×M[;1])∘.×M[1;]
[12]    →L×ι0≠I←I-1
[13]    Z←M[;⍋P]
    ∇
```

Figure 6.3

RLMATRIX forms the matrix G which will be used for displaying the estimate and true values of m by calling PLOT.

```
∇RLMATRIX[⎕]∇
     ∇ RLMATRIX
[1]    G←⍉(3,N)⍴(⍳N),M,EMA2
     ∇
```

Figure 6.4

Now the main part of the program is RLBEST, which lets us do the decision making. N is the number of points into which the interval (0,1) is reduced for the discretization (50 is a reasonable value). W is the weight vector in the moving average process, the choice of which will influence the regularity of the function m. One can start with something like 1,2,3,4,3,2,1. SIGN is the standard deviation and C is the constant in $m = C \cdot a \cdot (1-a) + n$. The program asks for the indices $a_1, a_2, \ldots$. The vector EMA1 is the vector that estimates m using the optimal form given in the text. EQ is the mean square error.

RLBEST is shown in figure 6.5.

```
        ∇RLBEST[⎕]∇
    ∇ RLBEST;I;C;A;NOISE1;CA;EMA1;J;RHA
[1]    'SPECIFY INTERVAL SUBDIVISION'
[2]    N←⎕
[3]    'SET WEIGHT VECTOR'
[4]    W←⎕
[5]    W←W÷(W+.×W)*0.5
[6]    'SET STANDARD DEVIATION'
[7]    SIGN←⎕
[8]    M←(W RLSERIES N)×SIGN
[9]    'SET C FOR M=C×A(1-A)'
[10]   C←⎕
[11]   F←C×A×1-A←(¯1+⍳N)÷(N-1)
[12]   M←M+F
[13]   T←OBS←⍳0
[14]   N←⍴M
[15]   R←N⍴0
[16]   I←0
[17]  LOOP:I←I+1
[18]   R[I]←W+.×((I⌊⍴W)↑W),(I⌊,⍴W)⍴0
[19]   →LOOP×⍳I<N
[20]   R←1,R
[21]   'SET OBS. POINT INDICES'
[22]   T←(,⎕)
[23]   P←⍴T
[24]   OBS←M[T]+NOISE1←(0 1) GAUSS P
[25]   EMA1←F+R[1+|(⍳N)∘.-T]+.×CA←(OBS-F[T])+.×RHA←INVP(R[1+|T∘.-T]+(⍳P)∘.=(⍳P))
[26]   'NUMBER OF OBS. MADE  ';P
[27]   'OBSERVATIONS WERE MADE AT INDICES:'
[28]   ⎕←T
[29]   MAXI←M⍳⌈/M
[30]   'MAX. OF MEAN VALUE FUNCTION OCCURRED AT:  ';MAXI
[31]   TOP1←EMA1⍳⌈/EMA1
[32]   'MAX. OF FIRST ESTIMATE OCCURRED AT:    ';TOP1
[33]   SUM←0
[34]   J←1
[35]   SUM←SUM+R[1+|T-J]+.×RHA+.×R[1+|T-J]
[36]   →35×⍳N≥J←J+1
[37]   ' MEASURE OF DEVIATION OF PRESENT ESTIMATE FROM TRUE VALUES M IS:'
[38]   ⎕←EQ←(SIGN*2)-((1÷N)×SUM)
    ∇
```

Figure 6.5

To go ahead and ask for more observations RLREST is called in.

```
          ∇RLREST[⎕]∇
      ∇ RLREST;NOISE2;CA;J
[1]     'SET ADDITIONAL OBS. POINTS'
[2]     TA←(,⎕)
[3]     Q←ρTA
[4]     OBS2←M[TA]+NOISE2←(0 1) GAUSS Q
[5]     T←T,TA
[6]     P←ρT
[7]     OBS←OBS,OBS2
[8]     EMA2←F+R[1+|(ιN(∘.-T]+.×CA←(OBS-F[T])+.×RHA←INVP(R[1+
        |T∘.-T]+(ιP)∘.=(ιP))
[9]     SUM←0
[10]    J←1
[11]    SUM←SUM+R[1+|T-J]+.×RHA+.×R[1+|T-J]
[12]    →11×ιN≥J←J+1
[13]    EQ2←(SIGN*2)-((1÷N)×SUM)
[14]    'NUMBER OF OBS. NOW MADE:   ';P
[15]    'OBSERV. WERE MADE AT INDICES:'
[16]    ⎕←T
[17]    TOP2←EMA2ι⌈/EMA2
[18]    'MAX. VALUE OF LATEST ESTIMATE OCCURRED AT INDEX:   ';
        TOP2
[19]    'MEASURE OF DEVIATION OF ESTIM. VECTOR SHOWS PREVIOUS
         VALUES AND LATEST:'
[20]    ⎕←EQ←EQ,EQ2
      ∇
```

Figure 6.6

The reader should use his subjective judgment to pick the a's wisely. He can also call RLSEL.

```
          ∇RLSEL[⎕]∇
      ∇ TA←RLSEL;PROB;Y;TA
[1]     PROB←50
[2]     Y←1?100
[3]     →5×ιY≤(100-PROB)
[4]     →7×ιY>(100-PROB)
[5]     TA←TOP2
[6]     →8×ιY≤(100-PROB)
[7]     TA←1?N
[8]     'SELECTED OBS.PT.   ';TA
      ∇
```

Figure 6.7

It chooses the observation point with a given probability PROB at the maximum of the last estimate and with the complementary probability picks a value at random from the whole interval.

RLSUM picks a point that maximizes EMA2 + B × DEV, where B is a given constant and DEV is the vector of variances of the estimate. EMA2 is the conditional expected value of the latest estimate.

```
        ∇RLSUM[⎕]∇
      ∇ TA←RLSUM;B;J;TA
[1]     B←2
[2]     DEV←⍳0
[3]     J←1
[4]     DEV←DEV,R[1+|T-J]+.×RHA+.×R[1+|T-J]
[5]     →4×⍳N≥J←J+1
[6]     VEC←EMA2+B×DEV←((SIGN*2)-DEV)*
        0.5
[7]     TA←VEC⍳⌈/VEC
[8]     'SELECTED OBS. PT.   ';TA
      ∇
```

Figure 6.8

```
RLBEST
SPECIFY INTERVAL SUBDIVISION
⎕:
      50
SET WEIGHT VECTOR
⎕:
      1 2 3 4 5 4 3 2 1
SET STANDARD DEVIATION
⎕:
      1
SET C FOR M=C×A(1-A)
⎕:
      20
SET OBS. POINT INDICES
⎕:
      1?50
NUMBER OF OBS. MADE  1
OBSERVATIONS WERE MADE AT INDICES:
49
MAX. OF MEAN VALUE FUNCTION OCCURRED AT:   32
MAX. OF FIRST ESTIMATE OCCURRED AT:   25
 MEASURE OF DEVIATION OF PRESENT ESTIMATE FROM TRUE VALUES M
       IS:
0.9597

      RLREST
SET ADDITIONAL OBSERVATION POINTS
⎕:
      1?50
NUMBER OF OBSERVATIONS NOW MADE:  2
OBSERVATIONS WERE MADE AT INDICES:
49  36
MAX. VALUE OF LATEST ESTIMATE OCCURRED AT INDEX:
      25
MEASURE OF DEVIATION OF ESTIMATE VECTOR SHOWS PREVIOUS VALUES
        AND LATEST:
0.9597  0.9069

      RLREST
SET ADDITIONAL OBSERVATION POINTS
⎕:
      RLSUM
SELECTED OBS. POINT  25
NUMBER OF OBSERVATIONS NOW MADE:  3
OBSERVATIONS WERE MADE AT INDICES:
49  36  25
MAX. VALUE OF LATEST ESTIMATE OCCURRED AT INDEX:
      25
MEASURE OF DEVIATION OF ESTIMATE VECTOR SHOWS PREVIOUS VALUES
        AND LATEST:
0.9597  0.9069  0.8541
```

Figure 6.9

```
      RLREST
SET ADDITIONAL OBSERVATION POINTS
⎕:
        RLSUM
     SELECTED OBS. POINT  29
     NUMBER OF OBSERVATIONS NOW MADE:  4
     OBSERVATIONS WERE MADE AT INDICES:
     49  36  25  29
     MAX. VALUE OF LATEST ESTIMATE OCCURRED AT INDEX:
           23
     MEASURE OF DEVIATION OF ESTIMATE VECTOR SHOWS PREVIOUS VALUES
            AND LATEST:
     0.9597  0.9069  0.8541  0.8114

           RLREST
     SET ADDITIONAL OBSERVATION POINTS
     ⎕:
           RLSUM
     SELECTED OBS. POINT  22
     NUMBER OF OBSERVATIONS NOW MADE:  5
     OBSERVATIONS WERE MADE AT INDICES:
     49  36  25  29  22
     MAX. VALUE OF LATEST ESTIMATE OCCURRED AT INDEX:
           26
     MEASURE OF DEVIATION OF ESTIMATE VECTOR SHOWS PREVIOUS VALUES
            AND LATEST:
     0.9597  0.9069  0.8541  0.8114  0.7752

           RLREST
     SET ADDITIONAL OBSERVATION POINTS
     ⎕:
           RLSUM
     SELECTED OBS. POINT  16
     NUMBER OF OBSERVATIONS NOW MADE:  6
     OBSERVATIONS WERE MADE AT INDICES:
     49  36  25  29  22  16
     MAX. VALUE OF LATEST ESTIMATE OCCURRED AT INDEX:
           26
     MEASURE OF DEVIATION OF ESTIMATE VECTOR SHOWS PREVIOUS VALUES
            AND LATEST:
     0.9597  0.9069  0.8541  0.8114  0.7752  0.7243

           RLREST
     SET ADDITIONAL OBSERVATION POINTS
     ⎕:
           RLSUM
     SELECTED OBS. POINT  27
     NUMBER OF OBSERVATIONS NOW MADE:  7
     OBSERVATIONS WERE MADE AT INDICES:
     49  36  25  29  22  16  27
     MAX. VALUE OF LATEST ESTIMATE OCCURRED AT INDEX:
           26
     MEASURE OF DEVIATION OF ESTIMATE VECTOR SHOWS PREVIOUS VALUES
            AND LATEST:
     0.9597  0.9069  0.8541  0.8114  0.7752  0.7243  0.7117
```

Figure 6.9 (continued)

```
      RLREST
SET ADDITIONAL OBSERVATION POINTS
⎕:
      RLSUM
SELECTED OBS. POINT  33
NUMBER OF OBSERVATIONS NOW MADE:  8
OBSERVATIONS WERE MADE AT INDICES:
49  36  25  29  22  16  27  33
MAX. VALUE OF LATEST ESTIMATE OCCURRED AT INDEX:
      26
MEASURE OF DEVIATION OF ESTIMATE VECTOR SHOWS PREVIOUS VALUES
       AND LATEST:
0.9597  0.9069  0.8541  0.8114  0.7752  0.7243  0.7117
      0.6847
```

Figure 6.9 (continued)

Figures 6.10 - 6.13 show the output of the above terminal session in graphical form.

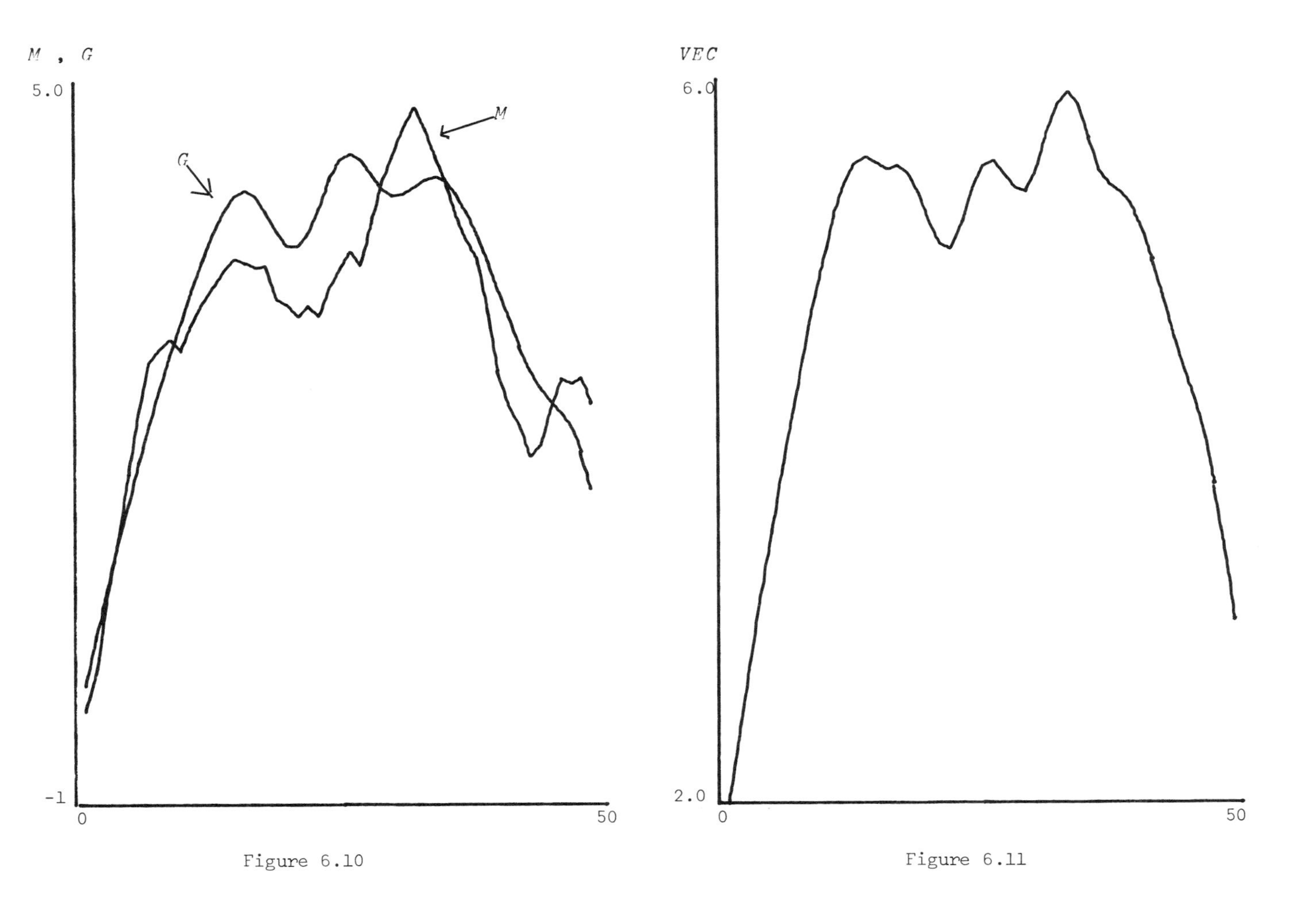

Figure 6.10

Figure 6.11

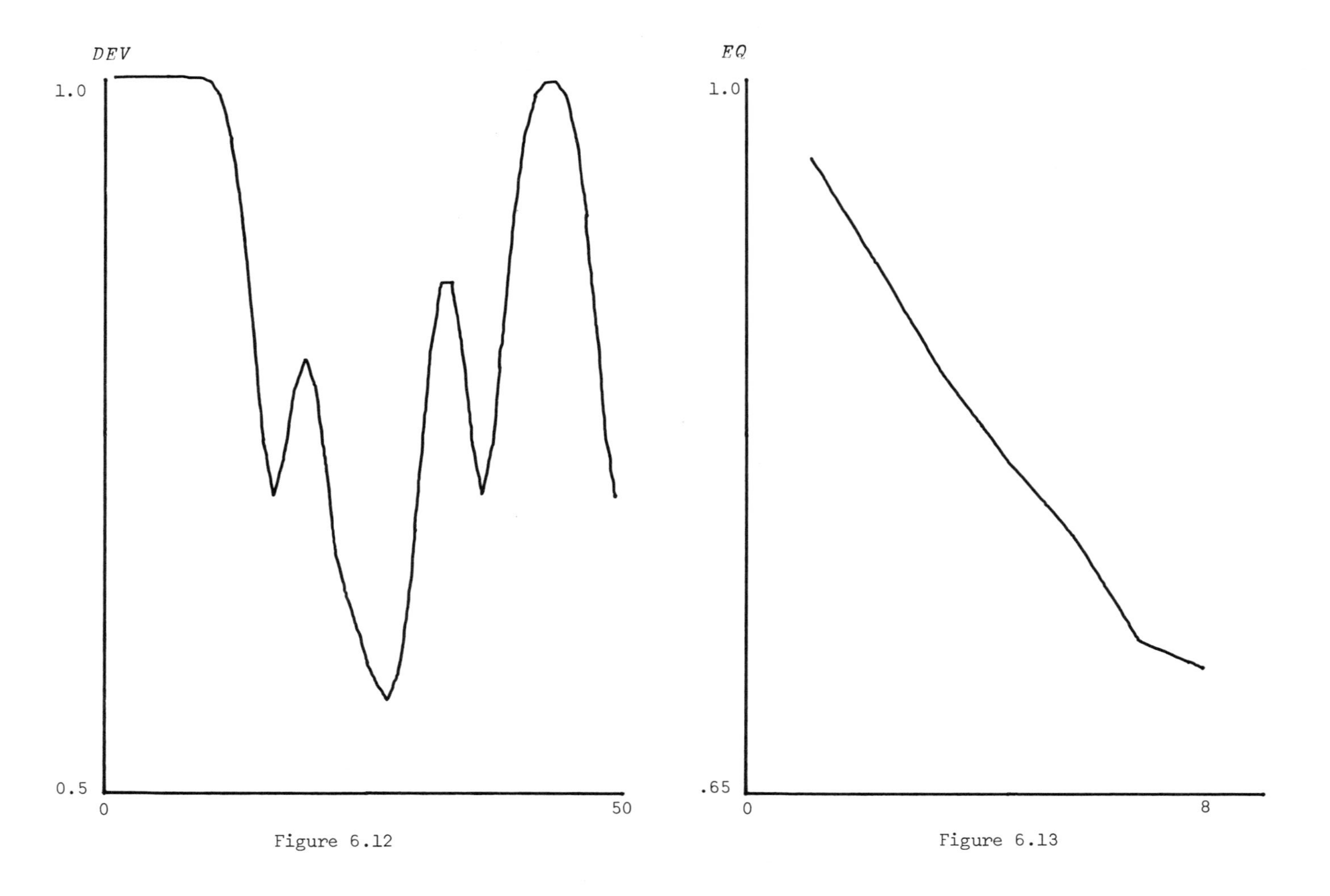

Figure 6.12

Figure 6.13

```
          ∇DESIGN[⎕]∇
     ∇ NUMBER DESIGN SS
[1]    I←1
[2]   DL:R←NUMBER[I]-1
[3]    (POWER+1) GRAM R
[4]    R+1
[5]    F←((¯1+⍳R+1)÷R)*POWER
[6]    (SS÷R+1)×(+/F×U[POWER+1;])*2
[7]    ⍳0
[8]    I←I+1
[9]    →(I≤⍴NUMBER)/DL
     ∇
```

```
          ∇GRAM[⎕]∇
     ∇ P GRAM R;R;I;V;M
[1]    I←2
[2]    U←(P,R+1)⍴0
[3]    U[1;]←(R+1)⍴÷(R+1)*I-1
[4]   GRAML:V←(¯1+⍳R+1)*0.5
[5]    M←U[⍳I-1;]
[6]    U[I;]←V-((⍉M)+.×M)+.×V
[7]    U[I;]←U[I;]÷(+/U[I;]*2)*0.5
[8]    I←I+1
[9]    →(I≤P)/GRAML
     ∇
```

Figure 6.14

```
        ∇LENGTH[⎕]∇
      ∇ Z←LENGTH L
[1]     →(L≤(N-1)÷2)/3
[2]     L←¯1+N-L
[3]     Z←0.5+0.5×0.9×0⌈1-L÷L0
      ∇
```

```
        ∇SEARCH[⎕]∇
      ∇ SEARCH P
[1]     N←ρP
[2]     CUMULATIVE←((ιN)∘.≥ιN)+.×P
[3]     I←1++/CUMULATIVE<0.001×?1000
[4]    SS1:'TYPE A'
[5]     A←⎕
[6]     'TYPE B'
[7]     B←⎕
[8]     TRUE←(0.001×?1000)≤LENGTH B-A
[9]     REL←(I≥A)∧I≤B
[10]    ('IO')[2-(TRUE∧REL)∨(1-TRUE)∧(1-REL)]
[11]    'TO CONTINUE TYPE 1'
[12]    CONT←⎕
[13]    →CONT=1)/SS1
      ∇
```

```
        ∇DECIDE[⎕]∇
      ∇ DECIDE P;A;B;V1;V0;CRIT
[1]     N←ρP
[2]     CRIT←0
[3]     AMAX←BMAX←A←B←1
[4]    SD1:IN←(A≤ιN)∧B≥ιN
[5]     V1←(IN×LENGTH B-A)+(1-IN)×1-LENGTH B-A
[6]     Q1←V1×P
[7]     P1←+/Q1
[8]     Q1←Q1÷P1
[9]     V0←(IN×1-LENGTH B-A)+(1-IN)×LENGTH B-A
[10]    Q0←V0×P
[11]    P0←+/Q0
[12]    Q0←Q0÷P0
[13]    DISPL←(P1×Q1*2)+P0×Q0*2
[14]    CRITN←+/DISPL
[15]    →(CRITN≤CRIT)/SD2
[16]    CRIT←CRITN
[17]    AMAX←A
[18]    BMAX←B
[19]   SD2:B←B+1
[20]    →(B≤N)/SD1
[21]    B←A←A+1
[22]    →(A≤N)/SD1
      ∇
```

Figure 6.15

CHAPTER 7:

A COMPUTATIONAL APPROACH TO STATISTICS

7.1 Statistical Computing.

There has been much controversy over the question what computational arrangements are best suited to statistical computing. Many packages have been offered for this purpose, with the developers sometimes advocating their exclusive use. Opponents, in turn, have argued that the idea of a statistical package is unsound per se since it encourages routine application of standard analyses for hypothesis testing, estimating, and so on.

The controversy actually goes back to pre-computer days when calculating schemes were playing the role of programs. Such schemes can be found in many and particularly in old textbooks; some of these may still be useful if one takes into account that present computing devices have affected questions of numerical efficiency.

Let us look somewhat more closely into the question of packages versus tailor-made programs, and let us start with a quite common and standard example. Assume that we have performed a factorial design with a few factors, the same number of replications in all cells, and no missing values. We want to assume the usual linear hypotheses, such as in (7.2.1), taking normal distributions with equal variances. Our task is to test for significance of the A-factor; how should we organize the computing effort?

One path is that taken in the next section: to write our own program suited to the problem at hand. This is relatively easy and if we are going to do the analysis only a few times we do not have to worry much about programming efficiency. On the other hand, if we plan to use the program in large-scale production we must devote more attention to such efficiency. We will then almost automatically be led to write the program in a somewhat more general form so that it can also be used in some other typical situations likely to arise. We would then get programs of the sort illustrated by ANOVA in figure 7.4 of Appendix 7 (figure 7.3 gives a description of ANOVA). It comes from a fairly recent package written in APL, viz. STATPACK, developed by Professor Smillie at the University of Alberta and his coworkers (ref. 9). For situations of this type packages seem well motivated.

We shall not discuss here the merits of the different packages being offered at present. Their number is increasing and we mention only that the one most widely used is probably the FORTRAN-based BMD, developed by Professor Dixon and his co-workers at UCLA (ref. 11).

When we turn from statistical situations of a type met fairly often in practice the perspective changes gradually. If the set-up is close (but not identical) to a standard one it may still be worthwhile to take a package and make the modifications called for by the application. This assumes that the package is sufficiently open-ended so that changes are simple enough to make. If this is not the case we may prefer to write our own program in a high-level language to be chosen.

One could use a statistical programming language especially designed to do the sort of operations that are common in statistics. Two such languages are COSMOS (Console Oriented Statistical Matrix Operator System, ref. 10) and PSTAT (Princeton Statistical System, ref. 12). The basic entity in COSMOS is the matrix together with operators on matrices e.g. SWEEP (Gaussian elimination). The usual arithmetic/logical operations, as well as I/O commands and a number of statistical utilities, are available in it.

Alternatively one could use APL, interactive FORTRAN, PL/I, etc., depending upon availability in the computer installation. It also depends, of course, upon the implementation of the language. As far as APL is concerned, the versions used in this course have been APL/360 and, only occasionally, APL/1130. In the first of these the work-spaces had sizes 32K or 64K bytes, which was seldom a serious limitation for the type of computing done. For larger statistical jobs larger work-spaces would have to be provided, especially since our APL implementation did not allow file manipulations on disk. Bulk I/O was not possible in our environment, and another drawback was the impossibility of communicating, in APL, with graphical display devices such as the 2250. Some of these defects are likely to be overcome in the near future.

The choice of the statistical computing set-up will depend entirely on the question how standard the problem is, whether it is a one-shot job or will be repeated a great number of times. Much of what is discussed in this course is aimed at making the reader familiar with the attitude of using the computer to do exploratory mathematics. It is then natural to base the computing effort on an interactive system.

7.2 Analysis of Variance

Let us first look at a case of traditional statistical analysis; later on we shall turn to situations in which the standard approaches are not valid.

Assume that we can observe the stochastic variables y_{ij}, $i = 1,2,\ldots,r$ and $j = 1,2,3,\ldots,s$. The y's represent data from a two-factor experiment where the first factor, A, appears on r levels and the second factor, B, appears on s levels. A represents treatments while B represents subjects. We are mainly interested in the effect of A: is it significant? Each combination i,j is replicated A times. The total number of observations is thus n = r+s.

To begin with, we shall try the usual model:

$$y_{ijk} = m + a_i + b_j + c_{ij} + x_{ijk}, \qquad k = 1,2 \tag{7.2.1}$$

with $\sum_i a_i = \sum_j b_j = \sum_i c_{ij} = \sum_j c_{ij} = 0$ and where we assume that $x_{ijk} = N(0,\sigma^2)$ and that the x_{ijk} are independent of each other. This is the analysis of variance model which can be found in most textbooks on statistics.

The standard procedure is as follows. Decompose the total sum of squares

$$Q = \sum(y_{ijk} - \bar{y}_{\ldots})^2 \tag{7.2.2}$$

into a number of terms, as follows. First form the total mean

$$\bar{y}_{\ldots} = \frac{1}{n} \sum_{i,j,k} y_{ijk} ; \tag{7.2.3}$$

form the sum of squares for the A effect, with $\bar{y}_{i..} = \frac{1}{st} \sum_{j,k} y_{ijk}$, namely

$$Q_A = st \sum_i (\bar{y}_{i..} - \bar{y}_{\ldots})^2 ; \tag{7.2.4}$$

form the sum of squares for the B effect, with $\bar{y}_{.j.} = \frac{1}{rt} \sum_{i,k} y_{ijk}$, namely

$$Q_B = rt \sum_j (\bar{y}_{.j.} - \bar{y}_{\ldots})^2 ; \tag{7.2.5}$$

form the interaction between A and B, with $\bar{y}_{ij.} = \frac{1}{t} \sum_{i,k} y_{ijk}$, namely

(7.2.6) $$Q_{AB} = t \sum_{i,j} (\bar{y}_{ij.} - \bar{y}_{i..} - \bar{y}_{.j.} + \bar{y}_{...})^2 ;$$

finally, form the residual

(7.2.7) $$Q_R = \sum_{i,j,k} (\bar{y}_{ijk} - \bar{y}_{ij.})^2$$

so that

(7.2.8) $$Q = Q_A + Q_B + Q_{AB} + Q_R$$

(Verify this identity!) We get the analysis

factor	sum of squares	degrees of freedom	ratio
A effect	Q_A	r-1	$Q_A/(r-1)$
B effect	Q_B	s-1	$Q_B/(s-1)$
AB effect	Q_{AB}	(r-1)(s-1)	$Q_{AB}/(r-1)(s-1)$
residual	Q_R	rs(t-1)	$Q_R/rs(t-1)$
total	Q	n-1	Q/(n-1)

From this table we can form the F ratios to test the significance of the various effects (see figure 7.1, which uses the array capabilities of APL).

We shall take Y as generated from another program, namely MODEL (see figure 7.2). Discuss numerical results, especially the influence of the interaction term when AB is not zero, and how this is related to the generation of $\bar{Y}$; is it consistent with the assumptions?

7.3 Non-Standard Situations.

What do we do when the standard assumptions break down, for instance when we do not trust the assumptions of linearity, normality, independence, etc.? We shall look briefly at this important problem area in this section.

Two questions have to be looked into. First, validity. Assume we use a test, confidence interval, or some other statistical decision function based on the usual assumptions of the linear hypothesis. If departures from these assumptions occur, how valid are the probabilistic statements we usually make? The departures from standard assumptions could affect, for instance, the level of significance of a test, the confidence level, or the standard deviation of an estimate. The second question concerns efficiency. In the usual set-up we have access to tests or estimates which are "best" in some precise sense. If the assumptions do not apply, how much efficiency do we lose?

If the decision function of interest to us has a simple form, we may be able to calculate its probabilistic properties analytically in closed form or we may at least be able to develop approximations (for example large sample approximations). We will otherwise have to use simulation, with the notoriously low accuracy to which this procedure usually leads (see chapter 2 for methods of getting higher accuracy). We will in the present context seldom need high accuracy, so that simulation may be acceptable.

Assignment. Assume that we want to deal with the null hypothesis H_0:m=0 against the one-sided alternative hypothesis H_1:m>0 when we have a sample of observations from $N(m,\sigma)$ and use Student's t. This is perhaps the most discussed case of all in statistical testing. We know how to get the best test on the level α.

Now assume that H_0 is replaced by a distribution which is still symmetric around x = 0 but has non-zero kurtosis. Take the frequency function

$$(7.3.1) \qquad f(x) = \frac{1}{2\sqrt{2\pi}\sigma}[e^{-(x+a^2)/2\sigma^2} + e^{-(x-a)^2/2\sigma^2}]$$

Design and carry out a simulation experiment to find the true level of the usual test when f(x) is given as above. Do this for different combinations of a and σ and present the result in a compact form. What conclusions do you draw concerning the change in validity?

Do the same but for an f that exhibits skewness. Choose a model where f is given as a mixture of normal frequency functions. Design the simulation scheme and analyze the result.

This leads naturally to consideration of the validity and efficiency of nonparametric competitors to the standard statistical techniques. Using the same approach as above we could study, say, the sign test, Wilcoxon's test and the Wallis-Kruskal test. Note, however, that at least for the sign test this approach is quite unnecessary. We could instead work analytically, which is quite simple in this case: when the analytical approach is feasible it should, of course, be used in preference to simulation.

Assume instead that we want to estimate m in $N(m,\sigma)$. We know that the arithmetic mean $\bar{x}$ is an efficient estimate. On the other hand, we also know that if the distribution is not normal (but still symmetric, say) and has long tails $\bar{x}$ may be a poor estimate. Instead we could introduce the ordered sample $(x_{(1)}<x_{(2)}<x_{(3)}< \dots <x_{(n)})$ and use a trimmed mean

$$\bar{x}_\alpha = \frac{x_{(i)}+x_{(i+1)}+\dots+x_{(n-i+1)}}{n-2i+2}$$

where we have trimmed away a portion $\alpha = (2i-2)/n$. Find, by simulation, what the relative efficiency of $\bar{x}$ and x_α are for the normal and non-normal distributions. Analyze the results; what conclusions do you draw?

In the chapter on time series we shall study the loss of efficiency in linear estimation due to dependence.

7.4 Bayesian Estimation.

Let α be a real-valued parameter which should be estimated and for which we have an a priori frequency function $p(\alpha)$. We observe a sample $x_1,x_2,\dots,x_n$ from a frequency function $f(x;\alpha)$ and ask for the a posteriori frequency function q_n of α, or, in other words, the conditional frequency function of α given the values $x_1,x_2,\dots,x_n$. We then get

$$q_n = q_n(\alpha;x_1,x_2,\dots,x_n) = \frac{p(\alpha)\prod_1^n f(x_\nu;\alpha)}{\int_{-\infty}^{\infty} p(\alpha)\prod_1^n f(x_\nu;\alpha)d\alpha} \quad (7.4.1)$$

This is Bayes' theorem.

Assignment. To get some intuitive feeling for the behavior of q_n as n increases, pick a function p and a statistical model $f(x;\alpha_0)$, where α_0 is some given value, and discretize α on a sufficiently fine level. Simulate the values $x_1,x_2,\dots,x_n$ from this model, calculate the discrete analogue of (7.4.1) and display it using a plotter or APL plots.

Note that (7.4.1) defines a stochastic mapping from the function space appropriate to p to the function space of q. As n increases it will be observed that p_n tends to become more concentrated around the "true value" α_0 of the parameter α. This property of the stochastic mapping shows up quite dramatically in the plots.

What is the analytic reason for this behavior? Try to outline a proof for it using the observation that

$$\ln q_n = \text{const} + \ln p + \sum_1^n \ln f(x_\nu;\alpha)$$

so that we have to deal with a sum of independent stochastic variables $\ln f(x_\nu;\alpha)$.

As an example of how the plots can appear see the photograph below which was obtained by taking Polaroid pictures of a Tektronix display tube connected to the DATEL terminal by a TSP control unit.

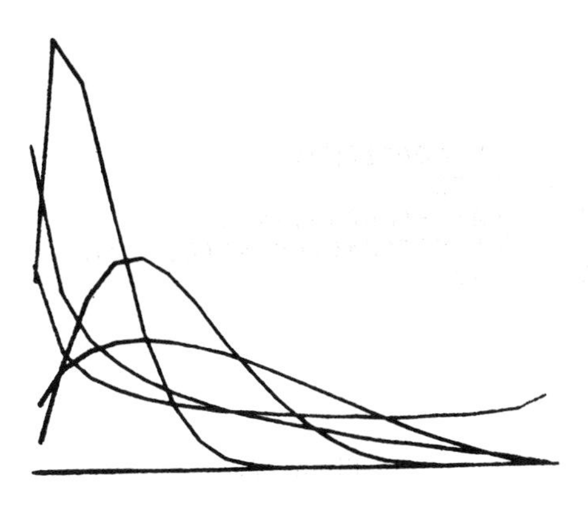

```
        ∇ ANOVA[⎕]∇
     ∇ ANOVA U
[1]    N←R×S×T
[2]    MA←(+/+/U)÷ST
[3]    MB←(+/[1]+/U)÷RT
[4]    MAB←(+/U)÷T
[5]    M←(+/+/+/U)÷N
[6]    Q←+/+/+/(U-M)*2
[7]    QA←S×T×+/(MA-M)*2
[8]    QB←R×T×+/(MB-M)*2
[9]    QAB←T×+/+/(MAB-(MA∘.×Sρ1)+((Rρ1)∘.×MB)-M)*2
[10]   QR←Q-QA+QB+QAB
[11]   'A RATIO IS ';QA÷R-1
[12]   'B RATIO IS ';QB÷S-1
[13]   'AB RATIO IS ';QAB÷(R-1)×(S-1)
[14]   ' RES RATIO IS ';QR÷R×S×T-1
[15]   'TOTAL RATIO IS ';Q÷N-1
     ∇
```

Figure 7.1

```
        ∇ MODEL[⎕]∇
     ∇ MODEL
[1]    Y←A∘.+(Sρ0)∘.+Tρ0
[2]    Y←Y+SIGMA×0.001×?(R,S,T)ρ1000
[3]    Y←*Y
     ∇
```

Figure 7.2

ANOVAHOW

COMPLETE FACTORIAL ANOVA
T←ANOVA D
ENTERED:25/10/67

THIS PROGRAM DOES AN ANALYSIS OF VARIANCE ON A COMPLETE FACTORIAL DESIGN WITH ARBITRARY NUMBERS OF REPLICATIONS AND FACTORS, WITH ARBITRARY NUMBERS OF LEVELS, WITH THE RESTRICTION THAT THERE ARE NO MISSING OBSERVATIONS.

T IS A MATRIX WITH 4 *COLUMNS FOR IDENTIFICATION, DEGREES OF FREEDOM, SUMS OF SQUARES, AND MEAN SQUARES. THE ROWS OF T REPRESENT REPLICATIONS, MAIN EFFECTS AND INTERACTIONS, ERROR AND TOTAL.*

AS AN EXAMPLE, CONSIDER THE FOLLOWING 2×2 *DESIGN WITH* 3 *REPLICATIONS*:

```
        1  2      5  6      9 10
        3  4      7  8     11 12
```

WHERE THE COLUMNS WITHIN EACH SQUARE REFER TO THE FIRST FACTOR A AND THE ROWS TO THE SECOND FACTOR B. THE DATA SHOULD BE PREPARED AS A VECTOR 1,2,....,12, *AND THEN RESTRUCTURED INTO D WITH DIMENSIONS* (3,2,2). *T WILL HAVE* 6 *ROWS FOR REPLICATIONS A,B AND AB EFFECTS, ERROR AND TOTAL. THE IDENTIFICATION IN THE FIRST COLUMN WILL BE* 1,10 *AND* 11 *FOR A,B AND AB, RESPECTIVELY, AND* 0*'S FOR THE REMAINING ROWS. IF IT IS DESIRED TO TREAT THE DESIGN AS A* 3×2×2 *FACTORIAL WITH A SINGLE REPLICATION, THEN D MUST BE RESTRUCTURED TO HAVE DIMENSIONS* (1,3,2,2). *T WILL THEN HAVE* 8 *ROWS FOR A,B,C, AB,AC,BC, ABC AND TOTAL. THE ROWS FOR REPLICATIONS AND ERROR WILL BE OMITTED.*

REQUIRES SS.

Figure 7.3

Description of ANOVA (figure 7.9).

```
      ∇ANOVA[⎕]∇
    ∇ T←ANOVA D;DIM;N;REPS;K;R;CT;V;I;S
[1]   N←(ρDIM←ρD)-1
[2]   T←((R←(2*N)+2×K←(REPS←DIM[1])≥2),4)ρ0
[3]   CT←((N+1)ρ0) SS D
[4]   T[R; 2 3]←((×/DIM)-1),((N+1)ρ1) SS D
[5]   →(REPS=1)/7
[6]   T[1; 2 3]←(REPS-1),((ι(N+1))≤1) SS D
[7]   D←+/[1] D
[8]   DIM←1↓DIM
[9]   V←⌊((2*(N+1)-ιN)∘.|ιS)÷(2*N-ιN)∘.×(S←(2*N)-1)ρ1
[10]  V[;ι(2*N)-1]←V[;(+/(X∘.>X)+((ιρX)∘.≥ιρX)∧X∘.=X)ιιρX←+/[1] V]
[11]  I←1
[12]  T[I+K; 2 3]←(×/((V[;I]=1)/DIM-1)),(V[;I] SS D)÷REPS
[13]  →((2*N)>I←I+1)/12
[14]  T[;3]←T[;3]-CT
[15]  →(N=1)/20
[16]  I←2
[17]  DV←(Kρ0),(X←(~(~CT)∨.∧S)∧(CT←V[;I])∨.∧S←V[;ιI-1]),(R-(I+K-1))ρ0
[18]  T[I+K;3]←T[I+K;3]-+/T[;3]×DV
[19]  →((2*N)>I←I+1)/17
[20]  →(REPS=1)/23
[21]  T[R-1;2]←T[R;2]-+/[1] T[ι(R-2);2]
[22]  T[R-1;3]←T[R;3]-+/[1] T[ι(R-2);3]
[23]  T[ι(R-1);4]←T[ι(R-1);3]÷T[ι(R-1);2]
[24]  I←1
[25]  T[I+K;1]←10⊥V[;I]
[26]  →((2*N)>I←I+1)/25
    ∇
```

Figure 7.4

From STATPACK (ref. 9).

CHAPTER 8:

TIME-SERIES ANALYSIS

8.1 Estimation of the Spectral Density.

A time series is a series of observations on a stochastic process. To analyze such a series it is necessary, first of all, to have a mathematical model - more or less well specified - which is thought to be appropriate for a description of the random process causing or underlying this sequence of observations.

Let us start with the model of random phase which we already discussed in section 5.2 and of which a simple form is

$$x_t = A\cos(\lambda t + \phi)\ ; \tag{8.1.1}$$

here λ is a known constant and ϕ is a random variable with rectangular distribution over $(0,2\pi)$. We are presented with a series of n observations x_t and wish to estimate the value of the unknown parameter A, or, rather, of A^2, the "power". To do this by Fourier analysis we form

$$\sum_{t=1}^{n} x_t \cos\lambda t = A\sum_{t=1}^{n} \cos\lambda t\cos(\lambda t + \phi) = \frac{A}{2}\sum_{t=1}^{n}\{\cos(2\lambda t + \phi) + \cos\phi\} \tag{8.1.2}$$

$$\sum_{t=1}^{n} x_t \sin\lambda t = A\sum_{t=1}^{n} \sin\lambda t\cos(\lambda t + \phi) = \frac{A}{2}\sum_{t=1}^{n}\{\sin(2\lambda t + \phi) - \sin\phi\} \tag{8.1.3}$$

The quantities $\sum_{t=1}^{n}\cos(2\lambda t + \phi)$ and $\sum_{t=1}^{n}\sin(2\lambda t + \phi)$ are each bounded as $n \to \infty$, which can be seen by writing

$$\sum_{t=1}^{n}\cos(2\lambda t + \phi) + i\sum_{t=1}^{n}\sin(2\lambda t + \phi) = \sum_{t=1}^{n} e^{i(2\lambda t+\phi)}$$

$$= \left|e^{i(2\lambda+\phi)}\frac{1-e^{in\lambda}}{1-e^{i\lambda}}\right| \le \frac{1}{|1-e^{i\lambda}|}$$

Also, $\sum_{t=1}^{n}\cos\phi = n\cos\phi$ and $\sum_{t=1}^{n}\sin\phi = n\sin\phi$, so that

(8.1.4) $$\left(\sum_{t=1}^{n} x_t \cos\lambda t\right)^2 + \left(\sum_{t=1}^{n} x_t \sin\lambda t\right)^2 \quad {}_{n \to \infty} \quad \frac{A^2n^2}{4}$$

and if we call

(8.1.5) $$\frac{1}{2\pi n}\left[\left(\sum_{t=1}^{n} x_t \cos\lambda t\right)^2 + \left(\sum_{t=1}^{n} x_t \sin\lambda t\right)^2\right] = I_n(\lambda)$$

the "periodogram" of the series of observations $x_1, x_2, \ldots, x_n$, it is clear that a reasonable way of estimating A^2, at least for large n, would be by means of the "statistic"

(8.1.6) $$\frac{8\pi}{n} I_n(\lambda)$$

A^2 can be looked upon as a measure of the "power" of the process

(8.1.7) $$\int_0^{2\pi} A^2 \cos^2(\lambda t+\phi)dt = A^2 \int_0^{2\pi} \cos^2(\lambda t+\phi)dt = \frac{\pi}{\lambda} A^2$$

Let us now look at some numerical properties of this estimate. We shall first generate a time series on the computer; next, we shall use the periodogram to estimate its spectrum (which is proportional to A^2) and demonstrate some of the properties of this estimate, which will turn out to be rather undesirable. To get a better estimate, we shall "smooth" (in a sense to be defined) the periodogram estimate and display the improved performance numerically. Finally, after having acquired some feeling for their numerical properties, we shall treat these estimates analytically.

The sequence of observations will be generated by the model of random phase

(8.1.8) $$x_t = \sum_{k=1}^{m} \sqrt{\frac{4\pi}{m} f\left(\frac{k}{m}\pi\right)} \cos\left(\frac{k}{m}\pi t + \phi_k\right) \quad \text{for } t = 1,2,\ldots,n$$

which is a sum of m terms of the form (8.1.1). To relate this to the notation of section 5.2 we observe that there we defined the spectral mass of the process x_t as

(8.1.9) $$F_k = A_k^2/4$$

In this chapter we use instead the spectral density f(x) defined as the derivative of the spectral mass

(8.1.10) $$f(x) = dF(x)/dx$$

or, in the discretized case,

$$F_k = \frac{\pi}{m} f(\frac{\pi}{m} x) \tag{8.1.11}$$

since then $dx = \pi/m$, $x = (\pi/m)k$. The "phases" ϕ_k are k random variables taken to be rectangularly distributed over the interval $(0,2\pi)$.

The APL program to generate the process x_t is the defined function RPHASE (see figure 8.1 of Appendix 8). This function will produce the n-component vector X. For sample size n (i.e. N) = 20, the following realization is an example of X produced:

¯0.04869987688	0.1647130654	1.034269864	0.2664539086
¯1.010349986	¯0.5016076709	¯0.0855910226	0.3912632526
0.3229693862	0.5665805442	0.07185708918	0.4651755559
¯0.1170016198	¯0.4749518966	¯0.2097731811	¯0.8531030046
0.3905699227	¯0.2788443918	¯0.0005277712274	
¯0.09340216699			

RPHASE requires as input a specific spectral density $f(\frac{k}{m}\pi)$, denoted by F in the program. Several special cases of F were considered: a triangular spectral density, produced by the APL program SPECTRUM (figure 8.2) and the density corresponding to an autoregressive scheme

$$x_{t+1} - \rho x_t = \varepsilon_t \tag{8.1.12}$$

where ε_t are normally distributed random variables, $N(0,\sigma)$, with spectral density

$$f_\varepsilon(\lambda) = \sigma^2/2\pi \tag{8.1.13}$$

so that the spectral density of the x_t process is

$$f_x(\lambda) = \frac{f_\varepsilon(\lambda)}{|e^{i\lambda}-\rho|^2} = \frac{\sigma^2}{2\pi(1-2\rho\cos\lambda + \rho^2)} ; \tag{8.1.14}$$

the latter is produced by the APL program AUTOSPECT (figure 8.3).

A realization of x_t with N = 30, M = 50 and rectangular spectral density F is shown in figure 8.10 (F ← (10⍴1),40⍴0).

We now take these computer-generated values of the process x_t, pretend we do not know the spectral density of x_t, and attempt to estimate it by various methods.

The most straightforward method is periodogram analysis. The program PERIODOGRAM (figure 8.5) computes EST, the vector of values of the periodogram estimate

$$\frac{1}{2\pi N}\{(\sum_{t=1}^{N} x_t \sin\frac{\pi k t}{M})^2 + (\sum_{t=1}^{N} x_t \cos\frac{\pi k t}{M})^2\} \tag{8.1.15}$$

Figure 8.12 shows the results of one such computation. The plotting program TIMEPL is shown in figure 8.11. The true spectral density F in this case corresponds to an autoregressive scheme and is produced by AUTOSPECT. It is clear from figure 8.12 that the periodogram is not a good estimate of the spectral density since it fluctuates wildly. We therefore proceed to "smooth" the periodogram by a weight-function ("window") W(k), which means using the estimate

$$\sum_{k=-m}^{m} I(\frac{k\pi}{m})\, W(k) \tag{8.1.16}$$

This is done by the APL program SMOOTH (figure 8.4) which produces the smoothed spectral estimate vector EST; lines [2] and [3] of that program require input of W(k) in the form of a function WINDOW which may, for instance, have rectangular shape

```
WINDOW ← 3ρ ÷ 6
```

Figure 8.13 presents a realization of this smoothing procedure, with the true spectral density F again shown. It should be remembered in observing these examples that the sample size used is unrealistically small. With the fast Fourier transform, to be discussed in section 8.2, more realistic examples could easily be handled.

Let us now turn to the analytic treatment of this smoothing process.

The smoothed periodogram can be represented by a quadratic form

$$Q = \sum_{\nu,\mu=1}^{n} x_\nu x_\mu q_{\nu\mu} \tag{8.1.17}$$

where the x_ν are random variables which, for simplicity, we later take to be normally distributed. We wish to find statistical properties such as the mean and the variance of this estimate Q.

In the following APL programs we denote the covariance matrix of the x_ν by COV and (without danger of confusion) the matrix $\{q_{\nu\mu}\}$ by Q_i.

The mean value of the estimate Q is given by

$$(8.1.18) \qquad EQ = \sum_{\nu,\mu} cov_{\mu\nu}\, q_{\nu\mu} = \sum_{\mu} (cov\ Q)_{\mu\mu} = tr\ cov\ Q$$

and the corresponding APL program by

(8.1.19)

```
A ← COV + . × Q
+/+/(A+.×A) × ((ιN)∘.=ιN)
```

The variance of Q is given by

$$(8.1.20) \quad Var\ Q = \sum_{\nu,\mu,\alpha,\beta} cov(x_\nu x_\mu, x_\alpha x_\beta)\cdot q_{\nu\mu} q_{\alpha\beta} = \sum_{\nu,\mu,\alpha,\beta} (r_{\nu\alpha} r_{\mu\beta} + r_{\nu\beta} r_{\mu\alpha}) q_{\nu\mu} q_{\alpha\beta}$$

$$= (COV\cdot Q)_{\alpha\mu}(COV\cdot Q)_{\mu\alpha} + (COV\cdot Q)_{\beta\mu}(COV\cdot Q)_{\mu\beta}$$

$$= tr\ A^2 + tr\ A^2 = 2\ tr\ A^2$$

where $A = COV\cdot Q$.

The APL program HERGLOTZ (figure 8.9) produces the output COV.

We now turn to the computation of the coefficients $q_{\nu\mu}$ in the quadratic form.

The periodogram $I(\lambda)$ for $\lambda = k\pi/m$ was given by

$$(8.1.21) \qquad I(\lambda) = \frac{1}{2\pi n} \{(\sum_t x_t \cos\lambda t)^2 + (\sum_t x_t \sin\lambda t)^2\}$$

so that the smoothed spectral estimate can be written

$$(8.1.22) \qquad f^* = \sum_k I(\frac{k\pi}{m})[TW(m+k) + TW(m+1-k)] \ ;$$

here TW stands for "translated window". To relate this notation for our expression for the smoothed spectral estimate to our expression for it in quadratic form, we note, by comparing coefficients of $x_\nu x_\mu$, that

$$(8.1.23) \quad q_{st} = \sum_k \{TW(m+k) + TW(m+1-k)\} \frac{1}{2\pi n} \cos\frac{k\pi}{m}t \cdot \cos\frac{k\pi}{m}s \cdot \sin\frac{k\pi}{m}t \cdot \sin\frac{k\pi}{m}s$$

$$= \sum_k \{TW(m+k) + TW(m+1-k)\} \frac{1}{2\pi n} \cos\frac{k\pi}{m}(t-s)$$

The APL program to compute the matrix $\{q_{st}\}$ is given by COMPQ (see figure 8.6), where the index I takes the place of the index k above.

The complete analysis is now given by the APL program ANALYSIS, which produces the mean MEAN and the standard deviation SD of the smoothed spectral estimate. This program is exhibited in figure 8.7.

Assignment. Using the above programs or modifications of them, experiment with different windows and compare the estimatesobtained with the true spectral density. In particular, investigate how the bandwidth (which is the opposite to resolution) affects bias and variance of the spectral density.

8.2 The Fast Fourier Transform.

The fast Fourier transform (FFT) is a computational algorithm which calculates the finite Fourier transform of a series of n data points in approximately $n \log_2 n$ operations, rather than in n^2 operations as required by traditional methods. It has made computation of spectral densities in terms of covariance functions of time series, and vice versa, feasible and economical in situations which were previously inaccessible.

The finite Fourier transform of x_j is defined as

$$(8.2.1) \qquad f(k) = \frac{1}{m} \sum_{j=0}^{m-1} x_j \exp\left(\frac{2\pi i j}{m} k\right)$$

where $i = \sqrt{-1}$, $j = 0,1,\ldots,m-1$, and x_j are a sequence of complex numbers. In this notation, the periodogram of these observations is given by

$$(8.2.2) \qquad I_k \equiv I\left(\frac{k\pi}{m}\right) = \frac{m}{2\pi}|f(k)|^2 = \frac{m}{2\pi} f(k)\,\overline{f(k)}$$

The observations x_j can be expressed as the inverse finite Fourier transform of f(k):

$$(8.2.3) \qquad x_j = \sum_{k=0}^{m-1} f(k) \exp\left(\frac{2\pi i j}{m} k\right)$$

We shall discuss the computation of this latter complex Fourier series; the equivalence with sine-cosine series is easily established. Direct computation of this series will clearly involve n operations, where each such operation consists of a complex multiplication and addition: n^2 elementary operations in all. The fast Fourier transform algorithm, which reduces the number of elementary operations to

$n \log_2 n$, will now be described.

Let x_j be a sampled (complex) variable and let us include the midpoints of our sampling intervals so as to get 2n points $j = 0,1,2,\ldots,2n-1$. Let the Fourier series of the even-indexed points be written

$$(8.2.4) \qquad x_{2j} = \sum_{k=0}^{n-1} f_1(k) \exp(\frac{2\pi ij}{n} k), \qquad j = 0,1,\ldots,n-1$$

and of the odd-indexed points be written

$$(8.2.5) \qquad x_{2j+1} = \sum_{k=0}^{n-1} f_2(k) \exp(\frac{2\pi ij}{n} k), \quad j = 0,1,\ldots,n-1$$

Then the series for the full 2n points becomes

$$(8.2.6) \qquad x_j = \sum_{k=0}^{2n-1} f(k) \exp(\frac{2\pi ij}{2n} k), \qquad j = 0,1,2,\ldots,2n-1$$

The idea of the algorithm is to compute $f(k)$ from $f_1(k)$ and $f_2(k)$.

Note that

$$[\exp(\frac{2\pi i}{2n})]^2 = \exp(\frac{2\pi i}{n})$$

We can write (8.2.6) separately for even and odd indices and get

$$(8.2.7) \qquad x_{2j} = \sum_{k=0}^{2n-1} f(k) \exp(\frac{2\pi ij}{n} k)$$

and

$$(8.2.8) \qquad x_{2j+1} = \sum_{k=0}^{2n-1} f(k) \exp(\frac{2\pi ij}{n} k) \exp(\frac{2\pi i}{2n} k), \quad j = 0,1,2,\ldots,n-1$$

If we separate the terms with $k \geq n$ and use $\exp(\frac{2\pi in}{n}) = 1$ and $\exp(\frac{2\pi in}{n}) = -1$, we get

$$\sum_{k=n}^{2n-1} f(k) \exp(\frac{2\pi ij}{n} k) = \sum_{k=0}^{n-1} f(n+k) \exp(\frac{2\pi ij}{n} k)$$

and

$$\sum_{k=n}^{2n-1} f(k) \exp(\frac{2\pi ij}{n} k) \exp(\frac{2\pi i}{2n} k) = -\sum_{k=0}^{n-1} f(n+k) \exp(\frac{2\pi ij}{n} k) \exp(\frac{2\pi i}{2n} k)$$

Substituting these in (8.2.7) and (8.2.8) and equating coefficients of like powers of $\exp(\frac{2\pi i}{n})$ with the coefficients of $f_1(k)$ and $f_2(k)$ in (8.2.4) and (8.2.5), we get

$$f_1(k) = f(k) = f(n+k) \ ; \qquad f_2(k) = [f(k) - f(n+k)] \exp(\frac{2\pi i}{n} k$$

which, solved for $f(k)$, gives

(8.2.9) $$f(k) = \frac{1}{2}\{f_1(k) + f_2(k)\exp[-\frac{2\pi i}{2n}k]\}$$

(8.2.10) $$f(n+k) = \frac{1}{2}\{f_1(k) - f_2(k)\exp[-\frac{2\pi i}{2n}k]\} \qquad k = 0,1,\dots,n-1$$

Eq. (8.2.9) states that the f(k) for the first half of the frequency range for sampling with interval $\frac{1}{2}\Delta t$ are the averages of the results of the two samplings on the odd and even indices. The effect of the multiplication by $\exp(-\frac{2\pi i}{2n}k)$ is to shift the sample on the odd indices by an amount $\frac{1}{2}\Delta t$ to the left. The new set of frequencies for $k = n,n+1,\dots,2n-1$ is the set of differences between the coefficients for the samplings on the even and odd indices.

It can be shown from (8.2.9) and (8.2.10) that the algorithm does indeed require (conservatively) $n \log_2 n$ operations (or complex multiply-adds).

The advantages of the FFT algorithm lie not only in the fact that it requires only $n \log_2 n$ instead of n^2 operations but also in that one can, during the calculation, observe whether the frequency amplitudes at the n- and 2n-point samplings agree. This gives a check that the sampling density is sufficient and that the calculations are correct.

The APL program in figure 8.8 performs the FFT algorithm. The following correspondences between the names of variables in the program and the above notation should be noted:

M corresponds to 2n

A1 corresponds to $f_1(k)$

A2 corresponds to $f_2(k)$

C and S are respectively the real and complex parts of the complex exponential. Z is the output of the program and X is the input vector.

8.3 Regression Analysis of Time Series

Consider the time series

(8.3.1) $$y_t = a\phi_t + x_t$$

where we now have a deterministic component $a\phi_t$ added to our stationary random process x_t. ϕ_t is the sequence of regression functions - they may be trigonometric functions, or given powers of t, for example. The problem is to estimate the regression parameter a in terms of observed values of y_t. We assume the spectrum (or covariance matrix) of x_t to be known. In general, the deterministic component will depend on more than one parameter, but this simple case suffices to illuminate the problem, which is, essentially, to determine the "best" estimate for the regression parameters and investigate its properties.

The analysis is much facilitated by carrying it out in vector and matrix notation. Let y be the (column) vector of observed $\{y_t\}$, $t = 1,\ldots,n$, x of the $\{x_t\}$ and ϕ of the $\{\phi_t\}$; then (8.3.1) can be written

$$y = a\phi + x \tag{8.3.2}$$

Let R be the (known) covariance matrix of the random n-vector x.

The straightforward way of estimating a would be by the method of least squares. This is obtained by minimizing the sum of the squares of the components of x, i.e. minimizing $x'x$, where x' is the transpose of x:

$$x = y - a\phi$$

$$x' = y' - a\phi'$$

$$x'x = (y'-a\phi')(y-a\phi) = y'y - a(\phi'y+y'\phi) + a^2\phi'\phi$$

The value a of a minimizing this expression is the one for which

$$\frac{d}{da}(x'x) = 0 = 2\hat{a}\phi'\phi - \phi'y - y'\phi$$

i.e.

$$\hat{a} = \frac{\phi'y + y'\phi}{2\phi'} = \frac{\phi'y}{\phi'\phi} = \frac{\phi'(a\phi+x)}{\phi'\phi} = a + \frac{\phi'x}{\phi'\phi} \tag{8.3.3}$$

$\hat{a}$ is, of course, a random variable with its own distribution. If we assume the components of x all have mean zero, we have

$$E\hat{a} = a$$

so that $\hat{a}$ is an unbiased estimate of a. Its variance is

(8.3.4) $$\text{Var } \hat{a} = E[(a-a)^2] = E[\frac{\phi' x}{\phi'\phi} \cdot \frac{x'\phi}{\phi'\phi}] = (\phi'\phi)^{-1} \phi' R\phi(\phi'\phi)^{-1}$$

since $R = E(xx')$. It can be shown that the unbiased estimator of a which is best in the sense that it minimizes the variance of the estimate is that given by

(8.3.5) $$\tilde{a} = (\phi' R^{-1}\phi)^{-1} \phi' R^{-1} y = a + (\phi' R^{-1}\phi)\phi' R^{-1} x$$

The variance of a is easily seen to be

(8.3.6) $$\text{Var } \tilde{a} = E[(\tilde{a}-a)^2] = (\phi' R^{-1}\phi)^{-1}$$

Comparing the least-squares estimate $\hat{a}$ (8.3.3) and the "Markov" estimate $\tilde{a}$ (8.3.5), we observe that even when the structure, i.e. the covariance matrix R, of the x-process is known (which is by no means always the case), the computation of a may present numerical difficulties since it involves the inversion of the large matrix R.

Assignment. Assume that x_t is generated by

$$x_{t+1} - \rho x_t = \varepsilon_t$$

where ε_t is N(0,1) and the spectral density of x_t is, therefore,

$$f_x(t) = \frac{1}{2\pi(1-2\rho \cos\lambda + \rho^2)}$$

If the regression relation is

$$y_t = at + x_t$$

find the least-squares and the Markov estimates for several values of $0 < \rho < 1$ and sample size n. For the inversion of R use the Toeplitz matrix inversion routine TINV (figure 4.10); see eq. (4.3.2) for the definition of a Toeplitz matrix.

8.4 Signal Detection.

Signal detection can be looked upon as a problem of hypothesis testing on stochastic processes.

Let y_t be an observation at time t of a random process; we wish to decide whether or not a signal s_t is present in that observation or whether we are observing noise, n_t, only. If the observations are made at time points $t = 1,2,\ldots,n$, these hypotheses may be written

$$H_0:\ y_t = n_t \qquad t = 1,2,\ldots,n$$

$$H_1:\ y_t = n_t + s_t\ , \qquad t = 1,2,\ldots,n$$

Let us assume that the noise is Gaussian with spectral density $f_n(\lambda)$ and that the signal is also Gaussian, with spectral density $f_s(\lambda)$. The above hypotheses can be written, if $f(\lambda)$ is the spectral density of y_t,

$$H_0:\ f(\lambda) = f_0(\lambda) = f_n(\lambda)$$

$$H_1:\ f(\lambda) = f_1(\lambda) = f_n(\lambda) + f_s(\lambda)$$

The spectral densities are additive since we assume n_t and s_t to be independent processes.

One can now define a critical region W in the sample space Ω of all possible realizations ω of y_t in the sense that

$$\text{if } \omega \in W,\ H_0 \text{ is rejected}$$

$$\text{if } \omega \notin W,\ H_0 \text{ is accepted}$$

The probability that H_0 be rejected though true is denoted by $P_0(W)$ and that it be accepted though false by $P_1(W^*)$; the power of the test, i.e. the probability of rejection of H_0 when false, is therefore

$$P_1(W) = 1 - P_1(W^*)$$

It can be shown that the most powerful critical region W_{MP} is given by the Neymann-Pearson test

$$W_{MP} = \{y|L(y) > c\}$$

where we denote by y the vector of observations $y_1, y_2, \ldots, y_n$ and by $L(y)$ the likelihood function

$$L(y) = \frac{p_1(y_1 \ldots y_n)}{p_0(y_1 \ldots y_n)}$$

Here $p_0(x)$ and $p_1(x)$ are the probability distributions induced on Ω by H_0 and H_1, respectively, and are given by

$$p_0(x) = \frac{1}{(2\pi)^{n/2}\sqrt{\det R_0}} \exp(-\frac{1}{2} x^T R_0^{-1} x)$$

$$p_1(x) = \frac{1}{(2\pi)^{n/2}\sqrt{\det R_1}} \exp(-\frac{1}{2} x^T R_1^{-1} x) \; ;$$

R_0 is the covariance matrix of the noise and R_1 that of the signal plus noise; these matrices are given by

$$R_0 = \frac{1}{2\pi} \int_{-\pi}^{\pi} e^{i(\nu-\mu)} f_0(\lambda)\, d\lambda \; ; \qquad \nu,\mu = 1,2,\ldots,n$$

and similarly for R_1. Hence

$$L(x) = K \exp[\frac{1}{2} x^T(R_0^{-1} - R_1^{-1})x]$$

The likelihood function is a monotonically increasing function of the quadratic form

$$x^T(R_0^{-1} - R_1^{-1})x = x^T Q x \; ,$$

say; hence

$$W_{MP} = \{x | x^T(R_0^{-1} - R_1^{-1})x > c\}$$

For white noise, $R_0 = I$, the unit matrix, and $f_n(\lambda) \equiv 1$; also

$$W_{MP} = \{x | x^T(I - R_1^{-1})x > c\} \; ;$$

the covariance matrix of the signal will be denoted by

$$S = R_1 - R_0 = R_1 - I = \{s_{\nu\mu}; \; \nu,\mu = 1,2,\ldots,n\}$$

Let us take, for the elements $s_{\nu\mu}$ of the covariance matrix S of the signal

$$s_{\nu\mu} = \frac{1}{2\pi} \int_{-\pi}^{\pi} \frac{1-\rho^2}{|1- e^{i\lambda}|^2} e^{i(\nu-\mu)\lambda}\, d\lambda = \rho^{|\nu-\mu|}$$

which corresponds to the spectral density

$$f_s(\lambda) = \frac{1-\rho^2}{|1-\rho e^{i\lambda}|^2}$$

the "Poisson kernel" (see ref. 21).

Assignment. Using the white noise process and the Poisson kernel process (with ρ = .8) for noise and signal, respectively, design a large sample test for presence and absence of a signal.

Hint: use the test statistic

$$T = x^T(R_0^{-1} - R_1^{-1})x = \sum w_{\nu\mu} x_\nu x_\mu = x^T W x$$

say; find the expected value and variance of T under the null hypothesis:

$$E_0 T = \sum w_{\nu\mu} r_{\nu\mu} = \text{tr } W R_0 = \text{tr}(I - R_1^{-1} R_0)$$

$$\text{Var}_0 T = \sum w_{\nu\mu} w_{\alpha\beta} (r_{\nu-\beta} r_{\mu-\alpha} + r_{\nu-\alpha} r_{\mu-\beta})$$

$$= 2\ \text{tr}(W R_0)^2 = 2\ \text{tr}(I - R_1^{-1} R_0)^2$$

The normalized test statistic is

$$\frac{T - E_0 T}{\sqrt{\text{Var}_0 T}}$$

```
      ∇RPHASE[⎕]∇
    ∇ RPHASE
[1]   FI←○0.02×?Mρ100
[2]   MAT←(÷M)×○(ιN)∘.×ιM
[3]   MAT←2○MAT+(Nρ1)∘.×FI
[4]   X←(2*0.5)×MAT+.×(○2×F÷M)*0.5
    ∇
```

Figure 8.1

```
      ∇AUTO[⎕]∇
    ∇ AUTO
[1]   X←(TRANS+N)ρ0
[2]   X[1]←0
[3]   I←1
[4]  AUL:I←I+1
[5]   GAUSS
[6]   X[I]←(R0×X] I-1])+SIGMA×EPSILON
[7]   →(I<N)/AUL
[8]   X←(-N)↑X
    ∇
```

```
      ∇SPECTRUM[⎕]∇
    ∇ SPECTRUM
[1]   F←1-(÷M)×ιM
    ∇
```

```
      ∇GAUSS[⎕]∇
    ∇ GAUSS
[1]   EPSILON←¯6+0.01×+/?12ρ100
    ∇
```

Figure 8.2

```
      ∇AUTOSPECT[⎕]∇
    ∇ AUTOSPECT
[1]   F←(SIGMA*2)÷○2×(1+R0*2)-2×R0×2○(÷M)×○ιM
    ∇
```

Figure 8.3

```
      ∇SMOOTH[⎕]∇
    ∇ SMOOTH
[1]   EST←(⌽EST),EST
[2]   WINDOW←WINDOW,(M-ρWINDOW)ρ0
[3]   WINDOW←(⌽WINDOW),WINDOW
[4]   I←1
[5]   POWER←Mρ0
[6]  SML:POWER[I]←+/(I⌽WINDOW)×EST
[7]   I←I+1
[8]   →(I≤M)/SML
[9]   EST←POWER
    ∇
```

Figure 8.4

```
        ∇PERIODOGRAM[⎕]∇
    ∇ PERIODOGRAM
[1]   MAT←1○(÷M)×○(⍳M)∘.×⍳M
[2]   V1←MAT+.×X
[3]   MAT←2○(÷M)×○(⍳M)∘.×⍳N
[4]   V2←MAT+.×X
[5]   EST←((V1*2)+V2*2)÷2×○N
    ∇
```

Figure 8.5

```
        ∇COMPQ[⎕]∇
    ∇ COMPQ;I
[1]   COL←Nρ0
[2]   I←1
[3]  COMPQL:COL←COL+(÷○2×N)×(TW[M+I]+TW[M+1-I])×2○(÷M)×○I×¯1+⍳N
[4]   I←I+1
[5]   →(I≤M)/COMPQL
[6]   Q←COL[1+|(⍳N)∘.-⍳N]
    ∇
```

Figure 8.6

```
        ∇ANALYSIS[⎕]∇
    ∇ ANALYSIS
[1]   SD←MEAN←Mρ0
[2]   K←1
[3]   WINDOW←WINDOW,(M-ρWINDOW)ρ0
[4]   WINDOW←(ϕWINDOW),WINDOW
[5]   ANALYSISL;TW←KϕWINDOW
[6]   COMPQ
[7]   A←Q+.×COV
[8]   MEAN[K]←+/+/A×((⍳N)∘.=⍳N)
[9]   A←Q+.×COV
[10]  SD[K]←2×+/+/A*2
[11]  SD[K]←SD[K]*0.5
[12]  K←K+1
[13]  →(K≤M)/ANALYSISL
    ∇
```

Figure 8.7

```
        ∇FFT[⎕]∇
    ∇ Z←FFT X;M;N;Z;P;A1;A2;C;S;VV
[1]   →KEEP×⍳2<M←⍴X
[2]   Z←(2 2)⍴(0.5×X[1]+X[2]),0,(0.5×X[1]-X[2]),0
[3]   →0
[4]  KEEP:A1←0.5×FFT X[¯1+P←2×⍳N←M÷2]
[5]   A2←0.5×FFT X[P]
[6]   C←2○VV←○(1-⍳N)÷N
[7]   S←1○VV
[8]   A2←⍉(2,N)⍴((A2[;1]×C)-(A2[;2]×S)),(A2[;2]×C)+A2[;1]×S
[9]   Z←((M,2)⍴A1)+(M,2)⍴(,A2),,-A2
    ∇
```

Figure 8.8

```
        ∇HERGLOTZ[⎕]∇
    ∇ HERGLOTZ
[1]   COVSEQ←○2×(÷M)×(2○(÷M)×○1×(¯1+⍳N)∘.×⍳M)+.×F
[2]   COV←COVSEQ[1+|(⍳N)∘.-⍳N]
    ∇
```

Figure 8.9

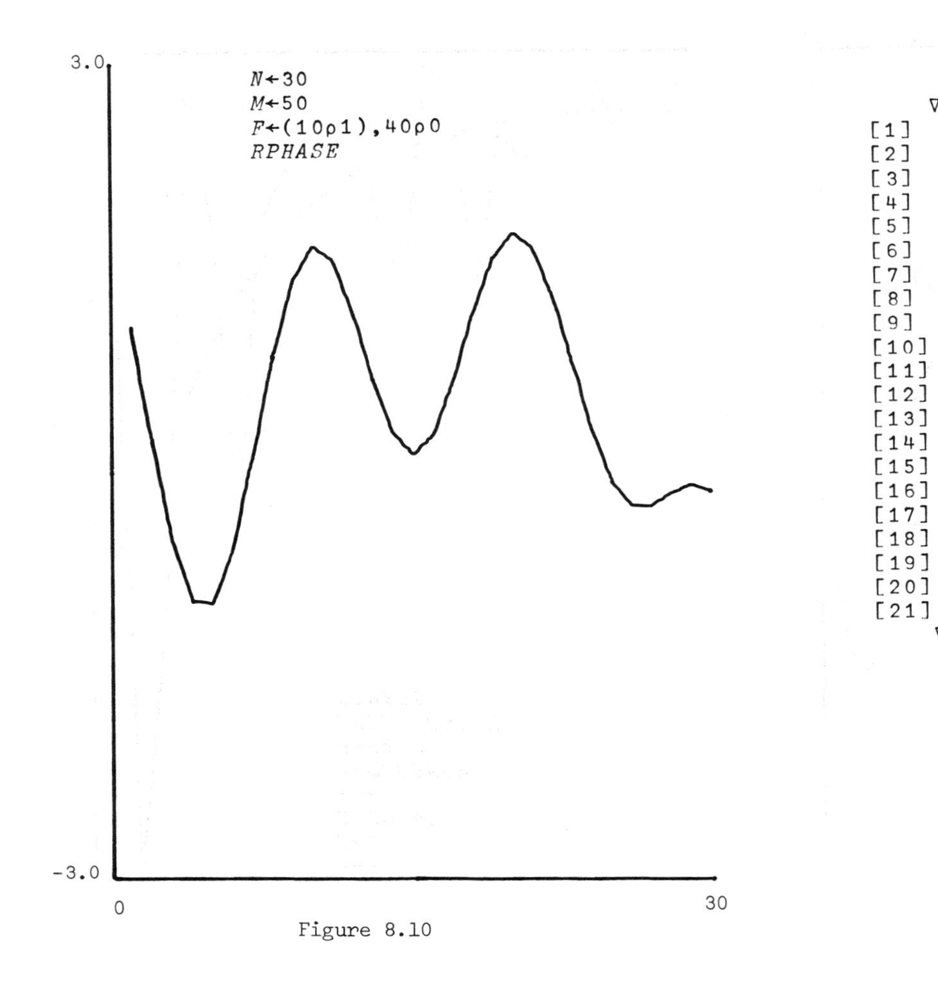

Figure 8.10

```
      ∇TIMEPL[⎕]∇
    ∇ TIMEPL
[1]   INITIALIZE
[2]   YMAX←⎕
[3]   INITIALIZE
[4]   OFFSET 0,5,0,0.1×YMAX
[5]   PLOT 0,0,13
[6]   PLOT 50,0,12
[7]   PLOT 0,0,13
[8]   PLOT 0,YMAX,12
[9]   T←⍳50
[10]  PLOT T[1],F[1],13
[11]  N←2
[12] LOOP:PLOT T[N],F[N],12
[13]  N←N+1
[14]  →(N≤50)/LOOP
[15]  TTY
[16]  PLOT T[1],EST[1],13
[17]  N←2
[18] LOOP2:PLOT T[N],EST[N],12
[19]  N←N+2
[20]  →(N≤50)/LOOP2
[21]  TTY
    ∇
```

Figure 8.11

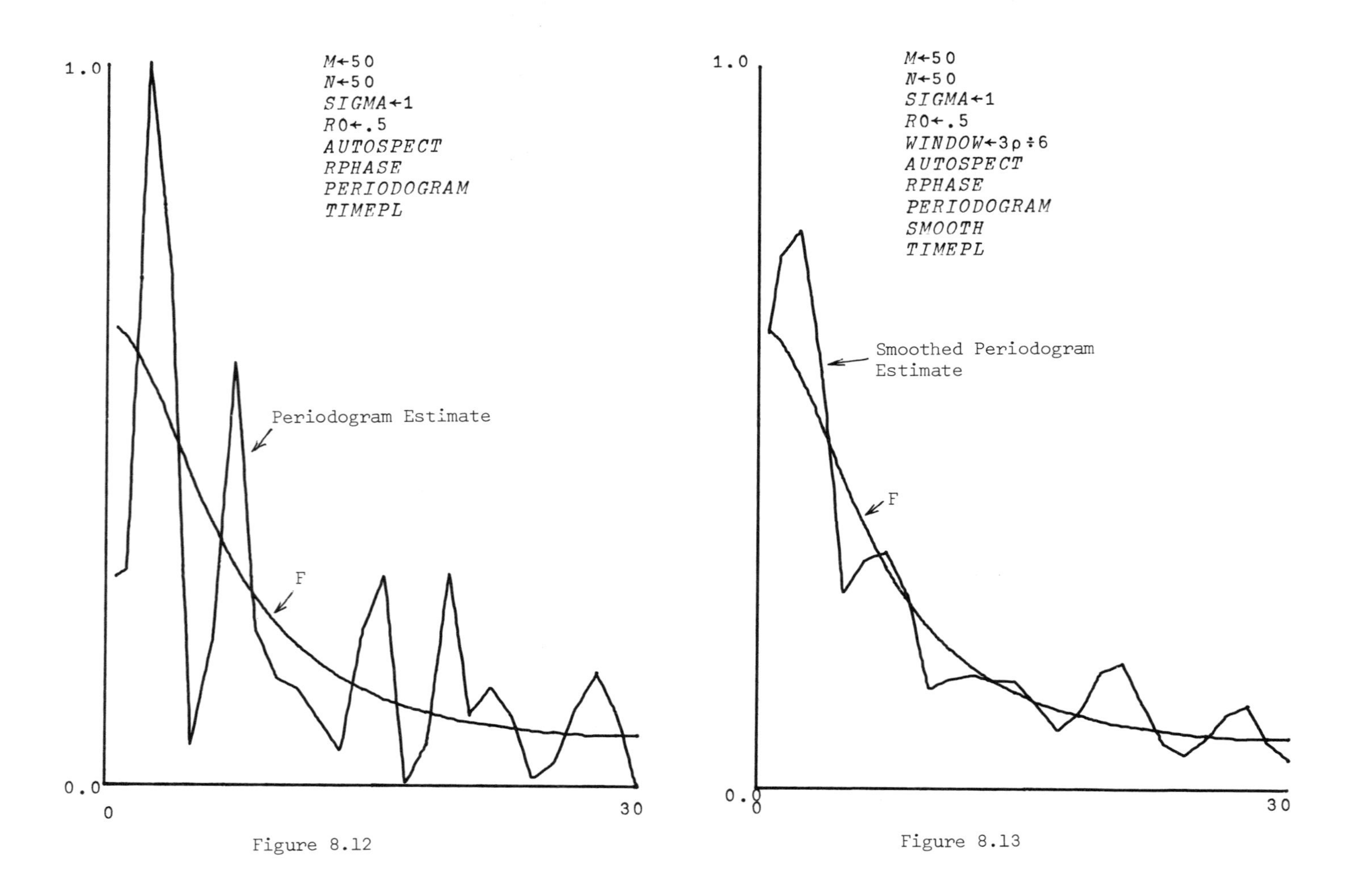

Figure 8.12

Figure 8.13

REFERENCES

1. B. Jansson: Random Number Generators. Almqvist & Wiksell, Stockholm (1966)

2. R.R. Coveyou and R.D. Macpherson: Fourier analysis of uniform random number generators, J. Assoc. Comp. Mach. 14: 100-119 (1967)

3. P.J. Davis and P. Rabinowitz: Numerical Integration. Ginn & Company, Boston (1967)

4. J.M. Hammersley and D.C. Handscomb: Proof of the antithetic variables theorem for n > 2, Proc. Camb. Philos. Soc. 54: 300-301 (1958)

5. G.H. Hardy, J.E. Littlewood and G. Pólya: Inequalities. Cambridge University Press, Cambridge (1934)

6. J.M. Hammersley and D.C. Handscomb: Monte Carlo Methods. John Wiley & Sons, Inc., New York (1964)

7. H. Cramér: Mathematical Methods of Statistics. Princeton University Press, Princeton (1946)

8. W. Feller: An Introduction to Probability Theory and its Applications, Vol. I, 3d edition. John Wiley & Sons, Inc., New York (1968)

9. K.W. Smillie: STATPACK 2: An APL statistical package. Second edition. Publication no. 17, Dept. of Computing Science, University of Alberta, Edmonton, Alberta (1969)

10. M. Schatzoff, R. Tsao and T. Burhoe: Design and implementation of COSMOS. IBM Cambridge Scientific Center Report (1967)

11. W.J. Dixon (editor): BMD - Biomedical Computer Programs. Health Sciences Computing Facility, School of Medicine, University of California, Los Angeles (1965)

12. R. Buhler et al.: P STAT - the Princeton Statistical System. Princeton University Computing Center, Princeton (1967)

13. M. Rosenblatt: Random Processes. Oxford University Press, New York (1962)

14. T.M. Apostol: Mathematical Analysis. Addison-Wesley Publishing Company, Inc. (1957)

15. IBM Scientific Subroutine Package Version III: Application, description, programmer's and system manuals.

16. W. Feller: An Introduction to Probability Theory and its Applications, Vol. II, 2d edition, John Wiley & Sons, Inc., New York (1971)

17. P. Martin-Löf: The definition of random sequences, Information and Control 9, 602-619 (1966).

18. B. Ajne and T. Dalenius: Some applications of statistical ideas to numerical integration (Swedish: English summary), Nordisk. Mat. Tidskr. 8, 145-152,198 (1960).

19. R.E. Beard, T. Pentikainen and E. Personen: Risk Theory. Methuen (1969)

20. U. Grenander and G. Szegö: Toeplitz Forms and their Applications. University of California Press (1958)

21. W. Freiberger: An approximate method in signal detection, Quart. Appl. Math. 20, 373-378 (1963)

22. L.H.C. Tippett: Random Sampling Numbers. Tracts for Computers 15, Cambridge University Press (1927).

Reports published to date under the Brown University "Computational Probability" project:

1. Outline
2. Randomness
3. Simulation
4. Regression in Time Series
5. Fortran Graphics Subroutines
6. Prediction
7. A Mutual Reinsurance Scheme
8. Training in Survey Sampling using Interactive Computing

INDEX

APPLIED MATHEMATICAL SCIENCES

Previously Published Volumes

Volume	ISBN	Title	Price
1	0-387-90021-7	Partial Differential Equations, F. John	$6.50
2	0-387-90022-5	Techniques of Asymptotic Analysis, L. Sirovich	$6.50
3	0-387-90023-3	Functional Differential Equations, J. Hale	$6.50
4	0-387-90027-6	Combinatorial Methods, J.K. Percus	$6.50
5	0-387-90028-4	Fluid Dynamics, K.O. Friedrichs and R. von Mises	$6.50
6		...er	$6.50